Student Solutions Manual

for

Thornton and Rex's

Modern Physics for Scientists and Engineers

Third Edition

Allen P. Flora
Hood College

BROOKS/COLE
CENGAGE Learning

Australia • Brazil • Japan • Korea • Mexico • Singapore • Spain • United Kingdom • United States

BROOKS/COLE
CENGAGE Learning™

**Student Solutions Manual
for Thornton and Rex's
Modern Physics for Scientists
and Engineers, Third Edition**
Allen P. Flora

For product information and
technology assistance, contact us at **Cengage Learning
Customer & Sales Support, 1-800-354-9706**

For permission to use material from this text or product,
submit all requests online at **cengage.com/permissions**
Further permissions questions can be emailed to
permissionrequest@cengage.com

ISBN-13: 978-0-534-41782-6

ISBN-10: 0-534-41782-5

Brooks/Cole
10 Davis Drive
Belmont, CA 94002-3098
USA

Cengage Learning is a leading provider of customized learning solutions with office locations around the globe, including Singapore, the United Kingdom, Australia, Mexico, Brazil, and Japan. Locate your local office at: **international.cengage.com/region**

Cengage Learning products are represented in Canada by Nelson Education, Ltd.

For your course and learning solutions, visit
academic.cengage.com

Purchase any of our products at your local college store or at our preferred online store **www.ichapters.com**

Printed in the United States of America
2 3 4 5 6   15 14 13 12 11

ED217

Preface and Acknowledgements

This book contains solutions to selected problems presented in *Modern Physics for Scientists and Engineers, Third Edition* by Thornton and Rex. Instructors who want students to have this manual, either as a required or an optional text, can order it through your bookstore (through Brooks/Cole Cengage Learning)

To the students using this manual, I want to remind you that there are many solutions to each particular physics problem. The one that you create is probably the one that you will understand best. However, the solutions presented here should provide enough detail to help clarify any problem with which you may have difficulty.

I would like to thank Stephen T. Thornton and Andrew Rex who offered suggested solutions for many of the problems and to Andrew Rex for allowing me to use solutions from earlier editions. I especially thank Donald Henry of Shepherd University who checked these solutions and offered many corrections and suggestions for clarification.

I welcome suggestions of alternate solutions and especially identification of any errors in this text, which are entirely mine.

Allen P. Flora
Department of Chemistry and Physics
401 Rosemont Avenue
Hood College
Frederick, MD 21701
flora@hood.edu

Table of Contents

Chapter 2

6. Let n = the number of fringes shifted; then $n = \dfrac{\Delta d}{\lambda}$. Since $\Delta d = c(\Delta t' - \Delta t)$, we have

$$n = \frac{c(\Delta t' - \Delta t)}{\lambda} = \frac{v^2(\ell_1 + \ell_2)}{c^2 \lambda}$$

Solving for v and noting that $\ell_1 + \ell_2 = 22$ m,

$$v = c\sqrt{\frac{n\lambda}{\ell_1 + \ell_2}} = \left(3.00 \times 10^8 \text{ m/s}\right)\sqrt{\frac{(0.005)(589 \times 10^{-9} \text{ m})}{22 \text{ m}}}$$

So $v = 3.47$ km/s.

7. Letting $\ell_1 \to \ell_1\sqrt{1 - \beta^2}$ (where $\beta = v/c$) the text equation for t_1 becomes

$$t_1 = \frac{2\ell_1\sqrt{1 - \beta^2}}{c(1 - \beta)} = \frac{2\ell_1}{c\sqrt{1 - \beta^2}}$$

which is identical to t_2 when $\ell_1 = \ell_2$, so $\Delta t = 0$ as required.

11. When $v \ll c$ we find $1 - \beta^2 \to 1$, so

$$x' = \frac{x - \beta ct}{\sqrt{1 - \beta^2}} \to x - \beta ct = x - vt$$

$$t' = \frac{t - \beta x/c}{\sqrt{1 - \beta^2}} \to t - \beta x/c \approx t$$

$$x = \frac{x' + \beta ct'}{\sqrt{1 - \beta^2}} \to x' + \beta ct' = x' + vt'$$

$$t = \frac{t' + \beta x'/c}{\sqrt{1 - \beta^2}} \to t' + \beta x'/c \approx t'$$

14. There is no motion in the transverse direction, so $y = z = 3.5$ m.

$$\gamma = \frac{1}{\sqrt{1 - \beta^2}} = \frac{1}{\sqrt{1 - 0.8^2}} = 5/3$$

$$x = \gamma\left(x' + vt'\right) = \frac{5}{3}\left(2\text{ m} + 0.8c\left(0\right)\right) = 10/3 \text{ m}$$

$$t = \gamma\left(t' + vx'/c^2\right) = \frac{5}{3}\left(0 + \left(0.8c\right)\left(2\text{ m}\right)/c^2\right) = 8.9 \times 10^{-9} \text{ s}$$

21. In the muon's frame $T_0 = 2.2\,\mu$s. In the lab frame the time is longer; see Equation (2.19): $T' = \gamma T_0$. In the lab the distance traveled is $9.5\text{ cm} = vT' = v\gamma T_0 = \beta c\gamma T_0$, since $v = \beta c$. Therefore

$$\beta = \frac{9.5\text{ cm}\left(\sqrt{1 - \beta^2}\right)}{cT_0}$$

So

$$\beta = \frac{v}{c} = \frac{9.5\text{ cm}\left(\sqrt{1 - \beta^2}\right)}{c\left(2.2\,\mu\text{s}\right)}$$

Now all quantities are known except β. Solving for β we find $\beta = 1.4 \times 10^{-4}$ or $v = 1.4 \times 10^{-4}\,c$.

26. a) $L' = L/\gamma = L\sqrt{1 - v^2/c^2} = 3.58 \times 10^4$ km $\sqrt{1 - 0.94^2} = 1.22 \times 10^4$ km

b) Earth's frame:

$$t = L/v = \frac{3.58 \times 10^7 \text{ m}}{\left(0.94\right)\left(3.00 \times 10^8 \text{ m/s}\right)} = 0.127 \text{ s}$$

Golf ball's frame:

$$t' = t/\gamma = 0.127 \text{ s } \sqrt{1 - 0.94^2} = 0.0433 \text{ s}$$

31. Start from the formula for velocity addition, Equation (2.23a):

$$u_x = \frac{u'_x + v}{1 + vu'_x/c^2}$$

a)

$$u_x = \frac{0.7c + 0.8c}{1 + (0.7c)(0.8c)/c^2} = \frac{1.5c}{1.56} = 0.96c$$

b)

$$u_x = \frac{-0.7c + 0.8c}{1 + (-0.7c)(0.8c)/c^2} = \frac{0.1c}{0.44} = 0.23c$$

36. We can ignore the 400 km, since it's small compared with the earth to moon distance of 3.84×10^8 m. The rotation rate is $\omega = 2\pi$ rad $\times$ $100\,\mathrm{s}^{-1} = 2\pi \times 10^2$ rad/s. Then the speed across the moon's surface is

$$v = \omega R = (2\pi \times 10^2 \text{ rad/s}) (3.84 \times 10^8 \text{ m}) = 2.41 \times 10^{11} \text{ m/s}$$

37. Classical:

$$t = \frac{4205 \text{ m}}{0.98c} = 1.43 \times 10^{-5} \text{ s}$$

Then

$$N = N_0 \exp\left[\frac{-(\ln 2)\, t}{t_{1/2}}\right] = 14.6 \text{ or about 15 muons}$$

Relativistic:

$$t' = t/\gamma = \frac{1.43 \times 10^{-5} \text{ s}}{5} = 2.86 \times 10^{-6} \text{ s}$$

$$N = N_0 \exp\left[\frac{-(\ln 2)\, t}{t_{1/2}}\right] = 2710 \text{ muons}$$

Because of the exponential nature of the decay curve, a factor of five (shorter) in time results in many more muons surviving.

42. For a timelike interval $\Delta s^2 < 0$ so $\Delta x^2 < c^2 \Delta t^2$. We will prove by contradiction. Suppose that there is a frame K' is which the two events were simultaneous, so that $\Delta t' = 0$. Then by the spacetime invariant

$$\Delta x^2 - c^2 \Delta t^2 = \Delta x'^2 - c^2 \Delta t'^2 = \Delta x'^2$$

But since $\Delta x^2 < c^2 \Delta t^2$, this implies $\Delta x'^2 < 0$ which is impossible because $\Delta x'$ is real.

48. The Doppler shift gives

$$\lambda = \lambda_0 \sqrt{\frac{1-\beta}{1+\beta}}$$

With numerical values $\lambda_0 = 670$ nm and $\lambda = 540$ nm, solving this equation for β gives $\beta = 0.212$. The astronaut's speed is $v = \beta c = 6.4 \times 10^7$ m/s. In addition to a red light violation, the astronaut gets a speeding ticket.

51.

$$f = f_0 \sqrt{\frac{1-\beta}{1+\beta}} = (400\ \text{Hz}) \sqrt{\frac{1-0.92}{1+0.92}} = 82\ \text{Hz}$$

54. The Doppler shift to higher wavelengths is (with $\lambda_0 = 589$ nm)

$$\lambda = 700\ \text{nm} \ = \lambda_0 \sqrt{\frac{1+\beta}{1-\beta}}$$

Solving for β we find $\beta = 0.171$. Then

$$t = \frac{v}{a} = \frac{(0.171)\,(3.00 \times 10^8\ \text{m/s})}{25\ \text{m/s}^2} = 2.052 \times 10^6\ \text{s}$$

which is 23.75 days. One problem with this analysis is that we have only computed the time as measured by earth. We are not prepared to handle the non-inertial frame of the spaceship.

55. Let the instantaneous momentum be in the x-direction and the force be in the y-direction. Then $d\vec{p} = \vec{F}\,dt = \gamma m\,d\vec{v}$ and $d\vec{v}$ is also in the y-direction. So we have

$$\vec{F} = \gamma m \frac{d\vec{v}}{dt} = \gamma m\,\vec{a}$$

60. The initial momentum is

$$p_0 = \gamma m v = \frac{1}{\sqrt{1-0.5^2}} m\,(0.5c) = 0.57735\,mc$$

Chapter 2

a) $p/p_0 = 1.01$

$$1.01 = \frac{\gamma m v}{0.57735\, mc}$$
$$\gamma v = (1.01)(0.57735\, c) = .58312\, c$$

Substituting for γ and solving for v,

$$v = \left[\frac{1}{(.58312\, c)^2} + \frac{1}{c^2}\right]^{-1/2} = 0.504\, c$$

b) Similarly

$$v = \left[\frac{1}{(.63509\, c)^2} + \frac{1}{c^2}\right]^{-1/2} = 0.536\, c$$

c) Similarly

$$v = \left[\frac{1}{(1.1547\, c)^2} + \frac{1}{c^2}\right]^{--/2} = 0.756\, c$$

67. a)

$$p = \gamma m u = \frac{(511\ \text{keV}/c^2)\,(0.0\llcorner c)}{\sqrt{1 - 0.01^2}} = 5.11\ \text{keV}/c$$

$$E = \gamma m c^2 = \frac{(511\ \text{keV}/c^2)\,(c^2)}{\sqrt{1 - 0.01^2}} = 511.03\ \text{keV}$$

$$K = E - E_0 = 511.03\ \text{keV} - 511.00\ \text{keV} = 30\ \text{eV}$$

The results for (b) and (c) follow with similar computations and are tabulated:

β	p (keV/c)	E (keV)	K (keV)
0.1	51.4	513.6	2.6
0.9	1055	1172	661

5

68. $E = 2E_0 = \gamma E_0$ so $\gamma = 2$. Then

$$\beta = \sqrt{1 - \frac{1}{\gamma^2}} = \frac{\sqrt{3}}{2}$$

and $v = \dfrac{\sqrt{3}\,c}{2}$.

74. It is the same for protons, electrons, or any particle.

$$K = (\gamma - 1)\,mc^2 = 1.01 \left(\frac{1}{2}mv^2\right) = 0.505mc^2\beta^2$$

$$\gamma - 1 = \frac{1}{\sqrt{1 - \beta^2}} - 1 = 0.505\beta^2$$

Rearranging and solving for β, we find $\beta = 0.114$ or $v = 0.114\,c$.

77. Up to Equation (2.57) the derivation in the text is complete. Then using the integration by parts formula

$$\int x\,dy = xy - \int y\,dx$$

and noting that in this case $x = u$ and $y = \gamma u$, we have

$$\int u\,d(\gamma u) = \gamma u^2 - \int \gamma u\,du$$

Thus

$$
\begin{aligned}
K &= m \int_0^{\gamma u} u\,d(\gamma u) = \gamma m u^2 - m \int \gamma u\,du \\
&= \gamma m u^2 - m \int \frac{u}{\sqrt{1 - u^2/c^2}}\,du
\end{aligned}
$$

Using integral tables or simple substitution

$$
\begin{aligned}
K &= \gamma m u^2 + mc^2 \sqrt{1 - u^2/c^2}\,\Big|_0^u \\
&= \gamma m u^2 + mc^2 \sqrt{1 - u^2/c^2} - mc^2 \\
&= \frac{mc^2 + mc^2\left(1 - u^2/c^2\right)}{\sqrt{1 - u^2/c^2}} - mc^2 = \gamma mc^2 - mc^2
\end{aligned}
$$

81.
$$E = K + E_0 = 1 \text{ TeV} + 938 \text{ MeV} \approx 1 \text{ TeV}$$

$$p = \frac{\sqrt{E^2 - E_0^2}}{c} = \frac{\sqrt{(1 \text{ TeV} + 938 \text{ MeV})^2 - (938 \text{ MeV})^2}}{c}$$

So $p = 1.000938 \text{ TeV}/c$.

$$\gamma = \frac{E + E_0}{E_0} = \frac{1.000938 \text{ TeV}}{0.000938 \text{ TeV}} = 1067$$

$$\beta^2 = 1 - \frac{1}{\gamma^2} = 1 - 8.78 \times 10^{-7}$$

$$\beta = \sqrt{1 - 8.78 \times 10^{-7}} \approx 1 - 4.39 \times 10^{-7}$$

$$v = \beta c \approx 0.999999561 \, c$$

83. $E = K + E_0 = 200 \text{ MeV} + 106 \text{ MeV} = 306 \text{ MeV}$

$$p = \frac{\sqrt{E^2 - E_0^2}}{c} = \frac{\sqrt{(306 \text{ MeV})^2 - (106 \text{ MeV})^2}}{c} = 287.05 \text{ MeV}/c$$

$$\gamma = \frac{E}{E_0} = \frac{306 \text{ MeV}}{106 \text{ MeV}} = 2.837$$

$$\beta = \sqrt{1 - \frac{1}{\gamma^2}} = 0.938 \quad \text{so} \quad v = 0.938 \, c$$

85. a) The mass-energy imbalance occurs because the helium-4 (^{4}He) is more tightly bound than the deuterium (^{2}H) and tritium nuclei (^{3}H).

$$
\begin{aligned}
\Delta E &= \left([m(^2\text{H}) + m(^3\text{H})] - [m_n + m(^4\text{He})] \right) c^2 \\
&= [(2.014102 \text{ u} + 3.016029 \text{ u}) - (1.008665 \text{ u} + 4.002603 \text{ u})] \, c^2 \\
&= (0.018863 \text{ u} \cdot c^2) \cdot \left(\frac{931.494 \text{ MeV}}{c^2 \cdot \text{u}} \right) \\
&= 17.6 \text{ MeV}
\end{aligned}
$$

b) The initial rest energy is $[m(^2\text{H}) + m(^3\text{H})] \cdot c^2$. This equals $(5.030131\,\text{u} \cdot c^2) \left(\dfrac{931.494\,\text{MeV}}{c^2 \cdot \text{u}} \right) = 4686\,\text{MeV}$. Thus the answer in a) is about 0.37 % of the initial rest energy.

87. As in the solution to Problem 21 we have

$$\beta = \frac{v}{c} = \frac{d\sqrt{1 - \beta^2}}{ct'}$$

where d is the length of the particle track and t' the particle's lifetime in its rest frame. In this problem $t' = 8.2 \times 10^{-11}$ s and $d = 24$ mm. Solving the above equation we find $\beta = 0.698$. Then

$$E = \frac{E_0}{\sqrt{1 - \beta^2}} = \frac{1672\,\text{MeV}}{\sqrt{1 - .698^2}} = 2330\,\text{MeV}$$

89. The number n received by Frank at f' is half the number sent by Mary at that rate, or $fL/\gamma v$. The detected time of turnaround is

$$t = \frac{n}{f'} = \frac{fL/\gamma v}{v\sqrt{(1 - \beta)/(1 + \beta)}} = \frac{L\sqrt{1 + \beta}}{\gamma v \sqrt{1 - \beta}} = \frac{L(1 + \beta)}{v} = \frac{L}{v} + \frac{L}{c}$$

Similarly, the number n' received by Mary at f' is

$$n' = f'\frac{T'}{2} = f\sqrt{\frac{1 - \beta}{1 + \beta}} \frac{L}{\gamma v} = \frac{fL(1 - \beta)}{v}$$

Her turnaround time is $T'/2 = L/\gamma v$.

100. As we know that the quasars are moving away at high speeds, we make use of Equation (2.33) and the equation $c = \lambda f$. Using a prime to indicate the Doppler shifted frequency (or wavelength), Equation (2.33) indicates that frequency is given by $f' = \dfrac{\sqrt{1 - \beta}}{\sqrt{1 + \beta}} f_0$

or $\dfrac{f_0}{f'} = \dfrac{\sqrt{1+\beta}}{\sqrt{1-\beta}}$

$$z = \frac{(\lambda' - \lambda_0)}{\lambda_0} = \left(\frac{\lambda'}{\lambda_0} - 1\right) = \left(\frac{c/f'}{c/f_0} - 1\right)$$

$$= \left(\frac{f_0}{f'} - 1\right) = \left(\frac{\sqrt{1+\beta}}{\sqrt{1-\beta}} - 1\right)$$

Therefore $(z+1)^2 = \dfrac{1+\beta}{1-\beta}$. We can complete the algebra to show that

$$\beta = \frac{(z+1)^2 - 1}{(z+1)^2 + 1} \quad \text{and thus } v = \left[\frac{(z+1)^2 - 1}{(z+1)^2 + 1}\right] c.$$

For the values of z given, $v = 0.787\,c$ for $z = 1.9$ and $v = 0.944\,c$ for $z = 4.9$.

Chapter 3

4. $eE = evB$ so $E = vB = (5.0 \times 10^6 \text{ m/s}) (1.3 \times 10^{-2} \text{ T}) = 6.50 \times 10^4 \text{ V/m}$

$$
\begin{aligned}
y &= \frac{1}{2}at^2 = \frac{1}{2}\left(\frac{F}{m}\right)\left(\frac{\ell}{v_0}\right)^2 = \frac{1}{2}\left(\frac{eE}{m}\right)\left(\frac{\ell}{v_0}\right)^2 = \frac{eE\ell^2}{2mv_0^2} \\
&= \frac{(1.602 \times 10^{-19} \text{ C}) (6.50 \times 10^4 \text{ V/m}) (0.02 \text{ m})^2}{2 (9.109 \times 10^{-31} \text{ kg}) (5.0 \times 10^6 \text{ m/s})^2} \\
&= 9.1452 \times 10^{-2} \text{ m} = 9.15 \text{ cm}
\end{aligned}
$$

7.
$$
v_t = \frac{mg}{f} = \frac{mg}{6\pi\eta r} \qquad\qquad m = \rho \, (\text{volume}) = \frac{4}{3}\pi\rho r^3
$$

$$
v_t = \left(\frac{4}{3}\pi\rho r^3\right)\left(\frac{g}{6\pi\eta r}\right) = \frac{2g\rho r^2}{9\eta}
$$

Solving for r

$$
r = 3\sqrt{\frac{\eta v_t}{2g\rho}}
$$

9. Lyman:

$$
\lambda = \left[R_H\left(1 - \frac{1}{\infty^2}\right)\right]^{-1} = R_H^{-1} = (1.096776 \times 10^7 \text{ m}^{-1})^{-1} = 91.2 \text{ nm}
$$

Balmer:

$$
\lambda = \left[R_H\left(\frac{1}{2^2} - \frac{1}{\infty^2}\right)\right]^{-1} = 4R_H^{-1} = 4\left(1.096776 \times 10^7 \text{ m}^{-1}\right)^{-1}
$$

$$
\lambda = 364.7 \text{ nm}
$$

13. Beginning with Equation (3.10) with $n = 1$, we have $\lambda = d\sin\theta$ with $d = 0.20\,\text{mm}$. Therefore $\dfrac{d\lambda}{d\theta} = d\cos\theta$. Assuming the spectra

10

was viewed in the forward direction, then $\theta \approx 0$ and $\cos \theta \approx 1$ so $\frac{d\lambda}{d\theta} \approx \frac{\Delta\lambda}{\Delta\theta} = d$. An angle of 0.50 minutes of arc corresponds to 1.45×10^{-4} rad so

$$\Delta\lambda = d\Delta\theta = (0.20 \times 10^{-3} \text{ m}) \, 1.45 \times 10^{-4} \text{ rad} = 29 \text{ nm}.$$

19.

$$\frac{P_1}{P_0} = \frac{\sigma T_1^4}{\sigma T_0^4} \quad \text{so} \quad P_1 = P_0 \frac{T_1^4}{T_0^4} = \left(\frac{1900 \text{ K}}{900 \text{ K}} \right)^4 P_0 = 19.9 \, P_0$$

The power increases by a factor of 19.9.

20. a)
$$\lambda_{\max} = \frac{2.898 \times 10^{-3} \text{ m} \cdot \text{K}}{310 \text{ K}} = 9.35 \, \mu\text{m}$$

b) At this temperature the power per unit area is

$$R = \sigma T^4 = (5.67 \times 10^{-8} \text{ W} \cdot \text{m}^{-2} \cdot \text{K}^4) \, (310 \text{ K})^4 = 524 \text{ W/m}^2$$

The total surface area of a cylinder is $2\pi r \, (r + h) = 2\pi \, (0.13 \text{ m}) \, (1.78 \text{ m})$ 1.45 m^2 so the total power is

$$P = \left(524 \text{ W/m}^2 \right) \left(1.45 \text{ m}^2 \right) = 760 \text{ W}.$$

23.
$$\lambda_{\max} = \frac{2.898 \times 10^{-3} \text{ m} \cdot \text{K}}{3000 \text{ K}} = 966 \text{ nm}$$

which is in the near infrared.

28. Taking derivatives

$$\frac{\partial^2 \psi}{\partial t^2} = \frac{\partial^2 a}{\partial t^2} \sin \left(\frac{n\pi x}{L} \right) \qquad \frac{\partial^2 \psi}{\partial x^2} = -a \frac{n^2 \pi^2}{L^2} \sin \left(\frac{n\pi x}{L} \right)$$

Substituting these values into the wave equation produces

$$\frac{1}{c^2} \frac{\partial^2 a}{\partial t^2} \sin \left(\frac{n\pi x}{L} \right) - \left(-a \frac{n^2 \pi^2}{L^2} \sin \left(\frac{n\pi x}{L} \right) \right) = 0$$

$$\frac{\partial^2 a}{\partial t^2} = -a\frac{n^2\pi^2 c^2}{L^2} = -\Omega^2 a$$

where $\Omega = \dfrac{n\pi c}{L}$. Since $\lambda = \dfrac{2L}{n}$ for this system and $c = \lambda f$, then $\Omega = 2\pi f$.

32.

$$\text{energy per photon} = hf = \left(6.626 \times 10^{-34} \text{ J} \cdot \text{s}\right)\left(107.7 \times 10^6 \text{ s}^{-1}\right)$$

$$\text{energy per photon} = 7.14 \times 10^{-26} \text{ J}$$

$$\left(5.0 \times 10^4 \text{ J/s}\right)\frac{1 \text{ photon}}{7.14 \times 10^{-26} \text{ J}} = 7.00 \times 10^{29} \text{ photons/s}$$

35.

$$\lambda_t = \frac{hc}{\phi} = \frac{1240 \text{ eV} \cdot \text{nm}}{4.64 \text{ eV}} = 267.2 \text{ nm}$$

If the wavelength is halved (to $\lambda = 133.6$ nm)

$$K = \frac{hc}{\lambda} - \phi = \frac{1240 \text{ eV} \cdot \text{nm}}{133.6 \text{ nm}} - 4.64 \text{ eV} = 4.64 \text{ eV}$$

43.

$$\lambda = \frac{1240 \text{ eV} \cdot \text{nm}}{2 \times 10^4 \text{ eV}} = 0.0620 \text{ nm}$$

45. From Figure (3.19) we observe that the two characteristic spectral lines occur at wavelengths of 6.4×10^{-11} m and 7.2×10^{-11} m. Rearrange Equation (3.37) to solve for the potential V_0;

$$V_0 = \frac{hc}{e}\frac{1}{\lambda_{min}} = \frac{1.24 \times 10^{-6} \text{ V} \cdot \text{m}}{7.2 \times 10^{-11} \text{ m}} = 17.2 \text{ kV}$$

and we have used the larger of the two wavelengths in order to determine the minimum potential.

50.

$$\frac{\Delta\lambda}{\lambda} = 0.004 = \frac{\lambda_c}{\lambda}(1 - \cos\theta) \qquad \text{so} \qquad \lambda = 250\lambda_c(1 - \cos\theta)$$

a)

$$\lambda = 250\left(2.43 \times 10^{-12} \text{ m}\right)(1 - \cos 30°) = 8.14 \times 10^{-11} \text{ m}$$

b)

$$\lambda = 250\left(2.43 \times 10^{-12} \text{ m}\right)(1 - \cos 90°) = 6.08 \times 10^{-10} \text{ m}$$

c)

$$\lambda = 250\left(2.43 \times 10^{-12} \text{ m}\right)(1 - \cos 170°) = 1.21 \times 10^{-9} \text{ m}$$

53. For $\theta = 90°$ we know $\lambda' = \lambda + \lambda_c = 2.00243$ nm

$$\frac{\Delta\lambda}{\lambda} = \frac{\lambda_c}{\lambda} = \frac{2.43 \times 10^{-3} \text{ nm}}{2 \text{ nm}} = 1.22 \times 10^{-3} = 0.122\%$$

54.

$$E = 2mc^2 = 2\,(938.3 \text{ MeV}) = 1877 \text{ MeV}$$

This energy could come from a particle accelerator.

58. For maximum recoil energy the scattering angle is $\theta = 180°$ and $\phi = 0$. Then as usual $\Delta\lambda = \frac{2h}{mc}$. Using the result of Problem 56

$$K = \frac{\Delta\lambda/\lambda}{1 + \Delta\lambda/\lambda}\,hf = \frac{2h/mc\lambda}{1 + 2h/mc\lambda}\,hf = \frac{2hf/mc^2}{1 + 2hf/mc^2}\,hf$$

For the given value of $K = 100$ keV we can solve this equation:

$$K\left(1 + \frac{2hf}{mc^2}\right) = \frac{2\,(hf)^2}{mc^2}$$

$$\left(\frac{2}{mc^2}\right)(hf)^2 - \left(\frac{2K}{mc^2}\right)(hf) - K = 0$$

This constitutes a quadratic equation in hf which can be solved numerically to yield $hf = 217$ keV.

Chapter 4

5. a) With $Z_1 = 2$, $Z_2 = 79$, and $\theta = 1°$ we have

$$b = \frac{Z_1 Z_2 e^2}{8\pi\epsilon_0 K} \cot\left(\frac{\theta}{2}\right) = \frac{(2)(79)(1.44 \times 10^{-9} \text{ eV} \cdot \text{m})}{2(7.7 \times 10^6 \text{ eV})} \cot(0.5°)$$
$$= 1.69 \times 10^{-12} \text{ m}$$

b) For $\theta = 90°$

$$b = \frac{Z_1 Z_2 e^2}{8\pi\epsilon_0 K} \cot\left(\frac{\theta}{2}\right) = \frac{(2)(79)(1.44 \times 10^{-9} \text{ eV} \cdot \text{m})}{2(7.7 \times 10^6 \text{ eV})} \cot(45°)$$
$$= 1.48 \times 10^{-14} \text{ m}$$

6.

$$f = \pi n t \left(\frac{Z_1 Z_2 e^2}{8\pi\epsilon_0 K}\right)^2 \cot^2\left(\frac{\theta}{2}\right)$$

For the two different angles everything is the same except the angles, so

$$\frac{f(1°)}{f(2°)} = \frac{\cot^2(0.5°)}{\cot^2(1.0°)} = 4.00$$

10. From the Rutherford scattering result, the number detected through a small angle is inversely proportional to $\sin^4\left(\frac{\theta}{2}\right)$. Thus

$$\frac{n(50°)}{n(6°)} = \frac{\sin^4(3°)}{\sin^4(25°)} = 2.35 \times 10^{-4}$$

and if they count 2000 at 6° the number counted at 50° is

$$(2000)\left(2.35 \times 10^{-4}\right) = 0.47$$

which is insufficient.

15. a)

$$v = \frac{e}{\sqrt{4\pi\epsilon_0 mr}} = \frac{ec}{\sqrt{4\pi\epsilon_0 mc^2 r}} = \frac{\sqrt{1.44 \text{ eV} \cdot \text{nm}}}{\sqrt{(938 \times 10^6 \text{ eV})(0.05 \text{ nm})}} c$$

$$= 1.75 \times 10^{-4} c = 5.25 \times 10^4 \text{ m/s}$$

b)

$$E = -\frac{e^2}{8\pi\epsilon_0 r} = -\frac{1.44 \text{ eV} \cdot \text{nm}}{2(0.05 \text{ nm})} = -14.4 \text{ eV}$$

c) The "nucleus" is too light to be fixed, and there is no way to reconcile this model with the results of Rutherford scattering.

22. From Equation (4.31) $v_n = (1/n)(\hbar/ma_0)$

$$n = 1: \quad v_1 = \frac{1}{1}\frac{1.055 \times 10^{-34} \text{ J} \cdot \text{s}}{(9.11 \times 10^{-31} \text{ kg})(5.29 \times 10^{-11} \text{ m})}$$

$$= 2.19 \times 10^6 \text{ m/s} = 0.0073c$$

$$n = 2: \quad v_2 = \frac{1}{2}\frac{1.055 \times 10^{-34} \text{ J} \cdot \text{s}}{(9.11 \times 10^{-31} \text{ kg})(5.29 \times 10^{-11} \text{ m})}$$

$$= 1.09 \times 10^6 \text{ m/s} = 0.0036c$$

$$n = 3: \quad v_3 = \frac{1}{3}\frac{1.055 \times 10^{-34} \text{ J} \cdot \text{s}}{(9.11 \times 10^{-31} \text{ kg})(5.29 \times 10^{-11} \text{ m})}$$

$$= 7.30 \times 10^5 \text{ m/s} = 0.0024c$$

23. The photon energy is

$$E = \frac{hc}{\lambda} = \frac{1240 \text{ eV} \cdot \text{nm}}{434 \text{ nm}} = 2.86 \text{ eV}$$

This is the energy difference between the two states in hydrogen. From Figure 4.16, we see $E_3 = -1.51$ eV so the initial state must be $n = 2$. We notice that this energy difference exists between $n = 2$ (with $E_2 = -3.40$ eV) and $n = 5$ (with $E_5 = -0.54$ eV).

30. The energy of each photon is $hc/\lambda = 12.4$ eV. Looking at the energy difference between levels in hydrogen we see that $E_2 - E_1 = 10.2$ eV, $E_3 - E_1 = 12.1$ eV, and $E_4 - E_1 = 12.8$ eV. There is enough energy to excite only to the second or third level. In theory it is possible for a second photon to come along and take the atom from one of these excited states to a higher one, but this is unlikely, because the $n = 2$ and $n = 3$ states are short-lived.

31. We must use the reduced mass for the muon:

$$\mu = \frac{mM}{m + M} = \frac{\left(106 \text{ MeV}/c^2\right)\left(938 \text{ MeV}/c^2\right)}{106 \text{ MeV}/c^2 + 938 \text{ MeV}/c^2} = 95.2 \text{ MeV}/c^2$$

a)

$$
\begin{aligned}
a_0 &= \frac{4\pi\varepsilon_0 \hbar^2}{\mu e^2} \\
&= \frac{\left(6.58 \times 10^{-16} \text{ eV} \cdot \text{s}\right)^2}{\left(1.44 \times 10^{-9} \text{ eV} \cdot \text{m}\right)\left(95.2 \times 10^6 \text{ eV}/c^2\right)} \frac{\left(3.00 \times 10^8 \text{ m/s}\right)^2}{c^2} \\
&= 2.84 \times 10^{-13} \text{ m}
\end{aligned}
$$

b)

$$E = \frac{e^2}{8\pi\varepsilon_0 a_0} = \frac{\left(1.44 \times 10^{-9} \text{ eV} \cdot \text{m}\right)}{2\left(2.84 \times 10^{-13} \text{ m}\right)} = 2535 \text{ eV}$$

c) First series:

$$\lambda = \frac{hc}{E} = \frac{1240 \text{ eV} \cdot \text{nm}}{2535 \text{ eV}} = 0.49 \text{ nm}$$

Second series:

$$\lambda = \frac{4hc}{E} = \frac{4\left(1240 \text{ eV} \cdot \text{nm}\right)}{2535 \text{ eV}} = 1.96 \text{ nm}$$

Third series:

$$\lambda = \frac{9hc}{E} = \frac{9\left(1240 \text{ eV} \cdot \text{nm}\right)}{2535 \text{ eV}} = 4.40 \text{ nm}$$

38. For L_α we have

$$\lambda = \frac{c}{f} = \frac{36}{5R(Z - 7.4)^2}$$

$Z = 43$: $\qquad \lambda = \dfrac{36}{5R(43 - 7.4)^2} = 0.52$ nm

$Z = 61$: $\qquad \lambda = \dfrac{36}{5R(61 - 7.4)^2} = 0.23$ nm

$Z = 75$: $\qquad \lambda = \dfrac{36}{5R(75 - 7.4)^2} = 0.14$ nm

41. Helium:

$$\lambda(K_\alpha) = \frac{4}{3R(Z - 1)^2} = 122 \text{ nm} \qquad \lambda(K_\beta) = \frac{9}{8R(Z - 1)^2} = 103 \text{ nm}$$

Lithium:

$$\lambda(K_\alpha) = \frac{4}{3R(Z - 1)^2} = 30.4 \text{ nm} \qquad \lambda(K_\beta) = \frac{9}{8R(Z - 1)^2} = 25.6 \text{ nm}$$

44. The longest wavelengths occur for an electron vacancy in K shell. The two longest wavelengths correspond to the K_α and the K_β. Use Equation (4.41) and rearrange it to solve for $(Z - 1)^2$. Then

$$(Z - 1)^2 = \frac{4}{3} \frac{1}{R \lambda_{K_\alpha}} = \frac{4}{3(1.09737 \times 10^7 \text{ m}^{-1})(0.155 \times 10^{-9} \text{ m})} = 783.89$$

This gives $(Z - 1) = 28$ or $Z = 29$. Using the expression for λ_{K_β} from problem 40, we have $(Z - 1)^2 = \dfrac{9}{8} \dfrac{1}{R \lambda_{K_\beta}}$. Using the second wavelength given, 0.131 nm, we find

$$(Z - 1)^2 = \frac{9}{8} \frac{1}{R \lambda_{K_\beta}} = \frac{9}{8(1.09737 \times 10^7 \text{ m}^{-1})(0.131 \times 10^{-9} \text{ m})} = 782.58$$

This yields $(Z - 1) = 27.97$ or $Z = 29$. Therefore we can conclude the target must be copper.

50. K_α is a transition from $n = 2$ to $n = 1$ and K_β is from $n = 3$ to $n = 1$. We know those wavelengths in the Lyman series are 121.6 nm and 102.6 nm, respectively. The redshift factor (λ/λ_0) is (with $\beta = 1/6$)

$$\sqrt{\frac{1+\beta}{1-\beta}} = \sqrt{\frac{1+1/6}{1-1/6}} = 1.183$$

Then the redshifted wavelengths are higher by 18.3% in each case. The observed wavelengths are:

$$K_\alpha: \ \lambda = (1.183)\,(121.6 \text{ nm}) = 143.9 \text{ nm}$$

$$K_\beta: \ \lambda = (1.183)\,(102.6 \text{ nm}) = 121.4 \text{ nm}$$

51.

$$f = \pi n t \left(\frac{Z_1 Z_2 e^2}{8\pi\varepsilon_0 K}\right)^2 \cot^2\left(\frac{\theta}{2}\right)$$

a)

$$f(1°) = \pi \left(5.90 \times 10^{28} \text{ m}^{-3}\right) \left(4 \times 10^{-7} \text{ m}\right)$$

$$\times \left(\frac{(2)\,(79)\,(1.44 \times 10^{-9} \text{ eV} \cdot \text{m})}{2\,(8 \times 10^6 \text{ eV})}\right)^2 \cot^2 (0.5°)$$

$$= 0.197$$

$$f(2°) = \pi \left(5.90 \times 10^{28} \text{ m}^{-3}\right) \left(4 \times 10^{-7} \text{ m}\right)$$

$$\left(\frac{(2)\,(79)\,(1.44 \times 10^{-9} \text{ eV} \cdot \text{m})}{2\,(8 \times 10^6 \text{ eV})}\right)^2 \cot^2 (1°)$$

$$= 0.0492$$

The fraction scattered between $1°$ and $2°$ is $0.197 - 0.0492 = 0.148$.

b)

$$\frac{f(1°)}{f(10°)} = \frac{\cot^2 (0.5°)}{\cot^2 (5°)} = 100.5$$

$$\frac{f(1°)}{f(90°)} = \frac{\cot^2 (0.5°)}{\cot^2 (45°)} = 1.31 \times 10^4$$

55. We start with $K = nhf_{orb}/2$. From classical mechanics we have for a circular orbit $f = v/2\pi r$, or $r = v/2\pi f$:

$$L = mvr = mv\left(\frac{v}{2\pi f}\right) = \left(\frac{mv^2}{2}\right)\left(\frac{1}{\pi f}\right)$$

Using $K = \frac{1}{2}mv^2$,

$$L = \frac{K}{\pi f} = \frac{nh\nu}{2\pi f} = \frac{nh}{2\pi} = n\hbar$$

Chapter 5

2. Use $\lambda = 0.186$ nm, and we know from the text that $d = 0.282$ nm for NaCl.

$$n = 1: \qquad \sin\theta = \frac{n\lambda}{2d} = \frac{\lambda}{2d} = 0.284 \qquad \theta = 19.3°$$

$$n = 2: \qquad \sin\theta = \frac{n\lambda}{2d} = \frac{\lambda}{d} = 0.567 \qquad \theta = 41.3°$$

$$\Delta\lambda = 41.3° - 19.3° = 22.0°$$

8. The resolution will be comparable to the de Broglie wavelength. The energy of the microscope requires a relativistic treatment, so

$$\lambda = \frac{h}{p} = \frac{hc}{\sqrt{K^2 + 2\,(mc^2)\,K}}$$

$$= \frac{1240\,\text{eV}\cdot\text{nm}}{\sqrt{(3 \times 10^6\,\text{eV})^2 + 2\,(0.511 \times 10^6\,\text{eV})\,(3 \times 10^6\,\text{eV})}}$$

$$= 3.57 \times 10^{-4}\,\text{nm} = 0.357\,\text{pm}$$

11. a) Relativistically

$$p = \frac{\sqrt{E^2 - E_0^2}}{c} = \frac{\sqrt{(K + mc^2)^2 - (mc^2)^2}}{c} = \frac{\sqrt{K^2 + 2Kmc^2}}{c}$$

$$\lambda = \frac{h}{p} = \frac{hc}{\sqrt{K^2 + 2Kmc^2}}$$

b) Non-relativistically, as in the text

$$\lambda = \frac{h}{\sqrt{2mK}} = \frac{hc}{\sqrt{2mc^2 K}}$$

12. The rest energy of the electron is very small compared to the total energy.

$$p = \frac{\sqrt{E^2 - E_0^2}}{c} = 50\,\text{GeV}/c$$

20

$$\lambda = \frac{h}{p} = \frac{1240 \text{ eV} \cdot \text{nm}}{50 \text{ GeV}} = 2.48 \times 10^{-17} \text{ m}$$

$$\text{fraction} = \frac{2.48 \times 10^{-17} \text{ m}}{2 \times 10^{-15} \text{ m}} = 0.012$$

15. From the accelerating potential we know $K = eV = 3$ keV.

$$E = K + E_0 = 514 \text{ keV}$$

$$p = \frac{\sqrt{E^2 - E_0^2}}{c} = \frac{\sqrt{(514 \text{ keV})^2 - (511 \text{ keV})^2}}{c} = 55.4 \text{ keV}/c$$

$$\lambda = \frac{h}{p} = \frac{hc}{pc} = \frac{1240 \text{ eV} \cdot \text{nm}}{55.4 \times 10^3 \text{ eV}} = 22.4 \text{ pm}$$

22.

$$\lambda = \frac{h}{\sqrt{2mK}} = \frac{hc}{\sqrt{2mc^2K}} = \frac{1240 \text{ eV} \cdot \text{nm}}{\sqrt{2\left(939 \times 10^6 \text{ eV}\right)\left(0.025 \text{ eV}\right)}}$$
$$= 0.181 \text{ nm}$$

$$\lambda = D \sin \phi \quad \phi = \sin^{-1}\left(\frac{\lambda}{D}\right) = \sin^{-1}\left(\frac{0.181 \text{ nm}}{0.45 \text{ nm}}\right) = 23.7°$$

29. As in Example 5.6 we start with $v_{ph} = c\lambda^n$ where c is a constant. We also know from Equation (5.33) that $u_g = v_{ph} + k\dfrac{dv_{ph}}{dk}$. We note further that $\dfrac{dv_{ph}}{dk} = \dfrac{dv_{ph}}{d\lambda}\left(\dfrac{d\lambda}{dk}\right)$. Since $\lambda = \dfrac{2\pi}{k}$ then $\dfrac{d\lambda}{dk} = \dfrac{-2\pi}{k^2} = \dfrac{-\lambda^2}{2\pi}$. Therefore

$$u_g = v_{ph} + \left(\frac{2\pi}{\lambda}\right)\frac{dv_{ph}}{d\lambda}\left(\frac{-\lambda^2}{2\pi}\right) = v_{ph} - (\lambda)\frac{dv_{ph}}{d\lambda}$$

Setting $u_g = v_{ph} = c\lambda^n$, we find $c\lambda^n = c\lambda^n - cn\lambda^n$. This can be satisfied only if $n = 0$, so v_{ph} is independent of λ. This is consistent with the idea that when a medium is non-dispersive, the phase and group velocities are equal and the speed independent of wavelength.

32. Relativistically

$$u = \frac{dE}{dp} = \frac{d}{dp}\left(p^2c^2 + E_0^2\right)^{1/2} = \frac{pc^2}{\sqrt{p^2c^2 + E_0^2}} = \frac{pc^2}{E}$$

Classically

$$u = \frac{dE}{dp} = \frac{d}{dp}\left(\frac{p^2}{2m}\right) = \frac{p}{m}$$

35. We make use of the small angle approximations: $\sin\theta \approx \tan\theta$ and $\sin\theta \approx \theta$.

$$\sin\theta \approx \tan\theta = \frac{0.3 \text{ mm}}{0.8 \text{ m}} = 3.75 \times 10^{-4}$$

$$\lambda = d\sin\theta \approx d\theta = (2000 \text{ nm})\left(3.75 \times 10^{-4}\right) = 0.75 \text{ nm}$$

$$p = \frac{h}{\lambda} = \frac{hc}{\lambda c} = \frac{1240 \text{ eV} \cdot \text{nm}}{(0.75 \text{ nm})\, c} = 1.653 \text{ keV}/c$$

$$
\begin{aligned}
K &= E - E_0 = \sqrt{p^2c^2 + E_0^2} - E_0 \\
&= \sqrt{(1.653 \text{ keV})^2 + (511 \text{ keV})^2} - 511 \text{ keV} = 2.67 \text{ eV}
\end{aligned}
$$

Such low energies will present problems, because low-energy electrons are more easily deflected by stray electric fields.

38. The uncertainty ratio is the same for any mass and independent of the box length.

$$\Delta p \Delta x = m \Delta v \Delta x \geq \frac{\hbar}{2} \qquad \text{or} \qquad \Delta v \geq \frac{\hbar}{2m\Delta x} = \frac{\hbar}{2mL}$$

$$E = \frac{1}{2}mv^2 = \frac{h^2}{8mL^2} \qquad \text{or} \qquad v = \sqrt{\frac{h^2}{4m^2L^2}} = \frac{h}{2mL}$$

$$\frac{\Delta v}{v} = \frac{\hbar/2mL}{h/2mL} = \frac{1}{2\pi}$$

43. a) $\Delta E \Delta t \geq \dfrac{\hbar}{2}$ so

$$\Delta E \geq \frac{\hbar}{2\Delta t} = \frac{6.582 \times 10^{-16} \text{ eV s}}{2 \left(1 \times 10^{-13} \text{ s}\right)} = 3.29 \times 10^{-3} \text{ eV}$$

b) Using the photon relation $E = \dfrac{hc}{\lambda}$ and taking a derivative

$$dE = -\frac{hc}{\lambda^2} d\lambda = -\frac{E^2}{hc} d\lambda$$

Then letting $\Delta\lambda = d\lambda$ and $\Delta E = dE$ we have

$$|\Delta\lambda| = hc\frac{\Delta E}{E^2} = (1240 \text{ eV} \cdot \text{nm}) \frac{3.29 \times 10^{-3} \text{ eV}}{(4.7 \text{ eV})^2} = 0.185 \text{ nm}$$

44. The wavelength of the electrons should be 0.14 nm or less. For this wavelength

$$p = \frac{h}{\lambda} = \frac{hc}{\lambda c} = \frac{1240 \text{ eV} \cdot \text{nm}}{(0.14 \text{ nm}) \, c} = 8.86 \text{ keV}/c$$

$$
\begin{aligned}
K &= E - E_0 = \sqrt{p^2 c^2 + E_0^2} - E_0 \\
&= \sqrt{(8.86 \text{ keV})^2 + (511 \text{ keV})^2} - 511 \text{ keV} = 77 \text{ eV}
\end{aligned}
$$

47. The proof is done in Example 6.11. With $\omega = \sqrt{k/m}$ we have a minimum energy

$$E = \frac{\hbar\omega}{2} = \frac{\hbar}{2}\sqrt{\frac{k}{m}} = \frac{1.055 \times 10^{-34} \text{ J} \cdot \text{s}}{2} \sqrt{\frac{8.2 \text{ N/m}}{0.0023 \text{ kg}}} = 3.15 \times 10^{-33} \text{ J}$$

59.

$$\Delta t = \frac{d}{c} = \frac{1.2 \times 10^{-15} \text{ m}}{3.0 \times 10^8 \text{ m/s}} = 4.0 \times 10^{-24} \text{ s}$$

$$\Delta E \geq \frac{\hbar}{2\Delta t} = \frac{6.582 \times 10^{-16} \text{ eV} \cdot \text{s}}{2 \left(4.0 \times 10^{-24} \text{ s}\right)} = 82 \text{ MeV}$$

This "lower bound" estimate of the rest mass is $\Delta E/c^2$ which is within a factor of two of the rest energy.

23

64. a) Starting from Equation (5.45), $\Delta E \Delta t \geq \dfrac{\hbar}{2}$ and substituting we have $\left(\dfrac{\Gamma}{2}\right)\tau \geq \dfrac{\hbar}{2}$. Therefore $\Gamma\tau \geq \hbar$.

b) We can find the minimum value for Γ from the equation above. Using the data for the neutron, for example,

$$\Gamma_{neutron} = \frac{6.582 \times 10^{-16}\,eV \cdot s}{887\,s} = 7.42 \times 10^{-19}\,eV.$$

The other values follow in a similar fashion.

$$\Gamma_{pion} = 2.53 \times 10^{-8}\,eV \qquad\qquad \Gamma_{upsilon} = 65.8\,keV$$

Chapter 6

4.

$$\Psi^*\Psi = A^2 \exp\left[-i(kx - \omega t) + i(kx - \omega t)\right] = A^2$$

$$\int_0^a \Psi^*\Psi dx = A^2 \int_0^a dx = A^2 a = 1 \qquad \text{so} \qquad A = \frac{1}{\sqrt{a}} \quad \text{and}$$

$$\Psi = \frac{1}{\sqrt{a}} \exp\left[i(kx - \omega t)\right]$$

5.

$$\Psi^*\Psi = A^2 r^2 \exp\left(\frac{-2r}{\alpha}\right)$$

$$\int_0^\infty \Psi^*\Psi dr = A^2 \int_0^\infty r^2 \exp\left(\frac{-2r}{\alpha}\right) dr = A^2 \left[\frac{2}{(2/\alpha)^3}\right] = \frac{A^2\alpha^3}{4} = 1$$

Therefore

$$A = \sqrt{\frac{4}{\alpha^3}} = 2\alpha^{-3/2}$$

10. Using the Euler relations between exponential and trig functions

$$\psi = A\left(e^{ix} + e^{-ix}\right) = 2A\cos(x)$$

Normalization:

$$\int_{-\pi}^{\pi} \psi^*\psi\, dx = 4A^2 \int_{-\pi}^{\pi} \cos^2(x)\, dx = 4A^2\pi = 1$$

Thus $A = \dfrac{1}{2\sqrt{\pi}}$ and the probability of being in the interval $[0, \pi/8]$ is

$$P = \int_0^{\pi/8} \psi^*\psi\, dx = \frac{1}{\pi}\int_0^{\pi/8} \cos^2(x)\, dx = \frac{1}{\pi}\left(\frac{x}{2} + \frac{1}{4}\sin(2x)\right)\Big|_0^{\pi/8}$$

$$= \frac{1}{16} + \frac{1}{4\pi\sqrt{2}} = 0.119$$

Chapter 6

13. The wave function for the nth level is $\psi_n(x) = \sqrt{\dfrac{2}{L}} \sin\left(\dfrac{n\pi x}{L}\right)$ so the average value of the square of the wave function is

$$
\begin{aligned}
\langle \psi_n^2(x) \rangle &= \frac{\int_0^L \psi^*\psi \, dx}{\int_0^L dx} = \frac{1}{L}\int_0^L \psi^*\psi \, dx = \frac{2}{L^2}\int_0^L \sin^2\left(\frac{n\pi x}{L}\right) dx \\
&= \frac{2}{L^2}\frac{L}{2} = \frac{1}{L}
\end{aligned}
$$

This result for the average value of the wave function is independent of n and is the same as the classical probability. The classical probability is uniform throughout the box, but this is not so in the quantum mechanical case which is $\dfrac{2}{L}\sin^2(k_n x)$.

15. a) Starting with Equation (6.35) and using the electron mass and the length given, we have

$$
\begin{aligned}
E_n &= n^2 \frac{\pi^2\hbar^2}{2mL^2} = n^2 \frac{\pi^2(\hbar c)^2}{2(mc^2)L^2} \\
&= n^2 \frac{\pi^2(197.3\,\text{eV}\cdot\text{nm})^2}{2(5.11\times10^5\,\text{eV})(2000\,\text{nm})^2} = n^2\left(9.40\times10^{-8}\,\text{eV}\right)
\end{aligned}
$$

Then the three lowest energy levels are: $E_1 = 9.40\times10^{-8}\,\text{eV}$; $E_2 = 1.88\times10^{-7}\,\text{eV}$; and $E_1 = 2.82\times10^{-7}\,\text{eV}$;

b) Average kinetic energy equals

$$
\frac{3}{2}kT = \frac{3}{2}\left[(1.381\times10^{-23}\,\text{J/K})\left(\frac{1\,\text{eV}}{1.602\times10^{-19}\,\text{J}}\right)\right]13\,\text{K}
$$

which equals $1.68\times10^{-3}\,\text{eV}$. Substitute this value into the equation above as E_n and solve for n. We find $n = 134$.

19.

$$
E_1 = \frac{h^2}{8mL^2} = \frac{h^2c^2}{8mc^2L^2} = \frac{(1240\,\text{eV}\cdot\text{nm})^2}{8(511\times10^3\,\text{eV})(10^{-5}\,\text{nm})^2} = 3.76\,\text{GeV}
$$

Then $E_2 = 4E_1 = 15.05\,\text{GeV}$ and $\Delta E = E_2 - E_1 = 11.3\,\text{GeV}$.

26

24. From the boundary condition $\psi_1(x = 0) = \psi_2(x = 0)$ we have $Ae^0 = Ce^0 + De^0$ or $A = C + D$. From the condition $\psi_1'(x = 0) = \psi_2'(x = 0)$ we have $\alpha A = ikC - ikD$. Solving this last expression for A and combining with the first boundary condition gives

$$C + D = \frac{ik}{\alpha} C - \frac{ik}{\alpha} D$$

or after rearranging

$$\frac{C}{D} = \frac{ik + \alpha}{ik - \alpha}$$

28. We must normalize by evaluating the triple integral of $\psi^*\psi$:

$$\iiint \psi^*\psi \, dxdydz = 1$$

with $\psi(x, y, z)$ given by Equation (6.47) in the text. We can evaluate the iterated triple integral

$$A^2 \int_0^L \sin^2\left(\frac{\pi x}{L}\right) dx \int_0^L \sin^2\left(\frac{\pi y}{L}\right) dy \int_0^L \sin^2\left(\frac{\pi z}{L}\right) dz = A^2 \left(\frac{L}{2}\right)^3 = 1$$

Solving for A we find

$$A = \left(\frac{2}{L}\right)^{3/2}$$

31.

$$\Delta E_n = E_{n+1} - E_n = \left(n + 1 + \frac{1}{2}\right)\hbar\omega - \left(n + \frac{1}{2}\right)\hbar\omega = \hbar\omega \quad \text{for all } n$$

This is true for all n, and there is no restriction on the number of levels.

36. Taking the derivatives for the Schrödinger equation

$$\frac{d\psi}{dx} = Ae^{-\alpha x^2/2} - A\alpha x^2 e^{-\alpha x^2/2}$$

27

$$\frac{d^2\psi}{dx^2} = -3A\alpha x e^{-\alpha x^2/2} + A\alpha^2 x^3 e^{-\alpha x^2/2} = \left(\alpha^2 x^2 - 3\alpha\right)\psi$$

Combining Equation (6.56) with this, we see

$$\frac{d^2\psi}{dx^2} = \left(\alpha^2 x^2 - \beta\right)\psi = \left(\alpha^2 x^2 - 3\alpha\right)\psi$$

Thus we see that $\beta = 3\alpha$ or

$$\frac{2mE}{\hbar^2} = 3\sqrt{\frac{mk}{\hbar^2}}$$

$$E = \frac{3}{2}\hbar\sqrt{\frac{k}{m}} = \frac{3}{2}\hbar\omega$$

40. In each case $\kappa L \gg 1$ so we can use

$$T = 16\frac{E}{V_0}\left(1 - \frac{E}{V_0}\right)e^{-2\kappa L}$$

where

$$\kappa = \frac{\sqrt{2mc^2\left(V_0 - E\right)}}{\hbar c} = \frac{\left(2\left(3727 \times 10^6 \text{ eV}\right)\left(10 \times 10^6 \text{ eV}\right)\right)^{1/2}}{197.4 \text{ eV} \cdot \text{nm}}$$
$$= 1.38 \times 10^{15} \text{ m}^{-1}$$

a) With $L = 1.3 \times 10^{-14}$ m

$$T_a = 16\frac{5 \text{ MeV}}{15 \text{ MeV}}\left(1 - \frac{5 \text{ MeV}}{15 \text{ MeV}}\right)e^{-2\left(1.38 \times 10^{15} \text{ m}^{-1}\right)\left(1.3 \times 10^{-14} \text{ m}\right)}$$
$$= 9.3 \times 10^{-16}$$

b) With $V_0 = 30$ MeV

$$\kappa = \frac{\sqrt{2mc^2\left(V_0 - E\right)}}{\hbar c} = \frac{\left(2\left(3727 \times 10^6 \text{ eV}\right)\left(25 \times 10^6 \text{ eV}\right)\right)^{1/2}}{197.4 \text{ eV} \cdot \text{nm}}$$
$$= 2.19 \times 10^{15} \text{ m}^{-1}$$

$$T_b = 16 \frac{5 \text{ MeV}}{30 \text{ MeV}} \left(1 - \frac{5 \text{ MeV}}{30 \text{ MeV}} \right) e^{-2(2.19 \times 10^{15} \text{ m}^{-1})(1.3 \times 10^{-14} \text{ m})}$$

$$= 4.2 \times 10^{-25}$$

c) With $V_0 = 15$ MeV we return to the original value of κ, but now $L = 2.6 \times 10^{-14}$ m and

$$T_c = 16 \frac{5 \text{ MeV}}{15 \text{ MeV}} \left(1 - \frac{5 \text{ MeV}}{15 \text{ MeV}} \right) e^{-2(1.38 \times 10^{15} \text{ m}^{-1})(2.6 \times 10^{-14} \text{ m})}$$

$$= 2.4 \times 10^{-31}$$

By comparison $T_a > T_b > T_c$.

47. As in the text we find

$$E = \frac{\hbar^2 \pi^2}{2m} \left(\frac{n_1^2}{L_1^2} + \frac{n_2^2}{L_2^2} + \frac{n_3^2}{L_3^2} \right)$$

and substituting the given values of L we find

$$E = \frac{\hbar^2 \pi^2}{2mL^2} \left(n_1^2 + 2n_2^2 + \frac{n_3^2}{4} \right)$$

Letting $E_0 = \hbar^2 \pi^2 / 2mL^2$ we have

$$E_1 = E_0 \left(1 + 2 + \frac{1}{4} \right) = \frac{13}{4} E_0$$

$$E_2 = E_0 \left(1 + 2 + \frac{2^2}{4} \right) = 4E_0$$

$$E_3 = E_0 \left(1 + 2 + \frac{3^2}{4} \right) = \frac{21}{4} E_0$$

$$E_4 = E_0 \left(2^2 + 2 + \frac{1}{4} \right) = \frac{25}{4} E_0$$

$$E_5 = E_0 \left(1 + 2 + \frac{4^2}{4} \right) = E_0 \left(2^2 + 2 + \frac{2^2}{4} \right) = 7E_0$$

Of those listed, only E_5 is degenerate.

49. a) In general inside the box we have a superposition of sine and cosine functions, but only the sine function satisfies the boundary condition $\psi(0) = 0$, and thus $\psi = A\sin(kx)$. With $V = 0$ inside the well $E = \dfrac{p^2}{2m} = \dfrac{\hbar^2 k^2}{2m}$ or $k = \dfrac{\sqrt{2mE}}{\hbar}$. Outside the well the decaying exponential is required as explained in section 6.4 of the text, with $E = \dfrac{\hbar^2 k^2}{2m} + V_0$ which reduces to $\kappa = ik = \dfrac{\sqrt{2m(V_0 - E)}}{\hbar}$.

b) Equating the wavefunctions and first derivatives at $x = L$:

$$A\sin(kL) = Be^{-\kappa L}$$

$$kA\cos(kL) = -\kappa Be^{-\kappa L}$$

Dividing these two equations

$$\frac{\tan(kL)}{k} = -\frac{1}{\kappa}$$

$$\kappa\tan(kL) = -k$$

57. The solution is identical to the presentation in the text for the three-dimensional box but without the z dimension. Briefly, we assume a trial function for the form

$$\psi(x, y) = A\sin(k_1 x)\sin(k_2 y)$$

Assuming that one corner is at the origin, applying the boundary conditions leads to

$$k_1 = \frac{n_x \pi}{L} \qquad\qquad k_2 = \frac{n_y \pi}{L}$$

and substituting into the Schrödinger equation leads to

$$E = \frac{\pi^2 \hbar^2}{2mL^2}\left(n_x^2 + n_y^2\right)$$

To normalize do the iterated double integral

$$\int_0^L \int_0^L \psi^* \psi \, dx \, dy \;=\; A^2 \int_0^L \int_0^L \sin^2\left(\frac{n_x \pi x}{L}\right) \sin^2\left(\frac{n_y \pi y}{L}\right) dx \, dy$$

$$=\; A^2 \left(\frac{L}{2}\right)\left(\frac{L}{2}\right) = 1$$

so $A = \dfrac{2}{L}$. Now to find the energy levels use the energy equation with different values of the quantum numbers. Letting $E_0 = \dfrac{\pi^2 \hbar^2}{2mL^2}$ we have

$$E_1 = E_0 \left(1^2 + 1^2\right) = 2E_0 \qquad\qquad n_1 = 1,\ n_2 = 1$$

$$E_2 = E_0 \left(2^2 + 1^2\right) = 5E_0 \qquad\qquad n_1 = 2,\ n_2 = 1 \text{ or vice versa}$$

$$E_3 = E_0 \left(2^2 + 2^2\right) = 8E_0 \qquad\qquad n_1 = 2,\ n_2 = 2$$

$$E_4 = E_0 \left(3^2 + 1^2\right) = 10E_0 \qquad\qquad n_1 = 3,\ n_2 = 1 \text{ or vice versa}$$

$$E_5 = E_0 \left(3^2 + 2^2\right) = 13E_0 \qquad\qquad n_1 = 3,\ n_2 = 2 \text{ or vice versa}$$

$$E_6 = E_0 \left(4^2 + 1^2\right) = 17E_0 \qquad\qquad n_1 = 4,\ n_2 = 1 \text{ or vice versa}$$

60. a) Note that this is an approximate procedure for one-dimensional problems with a gradually varying potential, $V(x)$. We begin with Equation (6.62b) which was derived for a scenario where $E > V$ but with V constant. We found $k = \dfrac{\sqrt{2m\left(E - V_0\right)}}{\hbar}$ for a constant V_0. Since our potential varies slowly, we approximate the wave number by $k = \dfrac{\sqrt{2m\left[E - V(x)\right]}}{\hbar}$. We also know that $p = \hbar k$ and $\lambda = \dfrac{h}{p}$. Combining all of the above, we find a position-dependent wavelength

$$\lambda(x) = \frac{h}{\sqrt{2m\left[E - V(x)\right]}}$$

b) If we neglect barrier penetration, then the wave function must be zero at the turning points. From the particle in a box example, we know that the number of wavelengths that fit between the turning points is $\frac{1}{2}$, or 1, or $\frac{3}{2}$, etc., which equals the distance divided by the wavelength. By analogy, the number of wavelengths that can fit inside our potential well with a slowly varying wavelength is

$$\int \frac{dx}{\lambda(x)} = \frac{n}{2} \; ; \text{ where } n \text{ is an integer.}$$

Substituting from above and rearranging, we have

$$2 \int \sqrt{2m\left[E - V(x)\right]}dx = nh \, ; \text{ where } n \text{ is an integer.}$$

Chapter 7

1. Starting with Equation (7.7), let the electron move in a circle of radius a in the xy-plane, so $\sin\theta = 1$. With both r and θ constant, R and f are also constant. Let $R = f = 1$. Then $g = \psi$ and the derivatives of R and f are zero. With this Equation (7.7) reduces to

$$-\frac{2\mu}{\hbar^2}a^2 (E - V) = \frac{1}{\psi}\frac{d^2\psi}{d\phi^2}$$

In uniform circular motion with an inverse-square force, we know from the planetary model that $E = V/2$, and

$$E - V = \frac{V}{2} - V = -\frac{V}{2} = |E|$$

Thus

$$-\frac{2\mu}{\hbar^2}a^2 |E| = \frac{1}{\psi}\frac{d^2\psi}{d\phi^2}$$

$$\frac{1}{a^2}\frac{d^2\psi}{d\phi^2} + \frac{2\mu}{\hbar^2}|E| = 0$$

5. Letting the constants in the front of R be called A we have

$$R = A\left(2 - \frac{r}{a_0}\right)e^{-r/2a_0}$$

$$\frac{dR}{dr} = A\left(-\frac{2}{a_0} + \frac{r}{2a_0^2}\right)e^{-r/2a_0}$$

$$\frac{d^2R}{dr^2} = A\left(\frac{3}{2a_0^2} - \frac{1}{4a_0^3}\right)e^{-r/2a_0}$$

Substituting these into Equation (7.13) we have

$$\left(-\frac{1}{4a_0^3} - \frac{2\mu E}{a_0\hbar^2}\right)r + \left(\frac{5}{2a_0^2} + \frac{4\mu E}{\hbar^2} - \frac{2\mu e^2}{4\pi\epsilon_0 a_0\hbar^2}\right)$$

$$+ \left(-\frac{4}{a_0} + \frac{4\mu e^2}{4\pi\epsilon_0\hbar^2}\right)\frac{1}{r} = 0$$

33

To satisfy the equation, each of the expressions in parentheses must equal zero. From the $1/r$ term we find

$$a_0 = \frac{4\pi\epsilon_0\hbar^2}{\mu e^2}$$

which is correct. From the r term we get

$$E = -\frac{\hbar^2}{8\mu a_0^2} = -\frac{E_0}{4}$$

which is consistent with the Bohr result. The other expression in parentheses also leads directly to $E = -E_0/4$, so the solution is verified.

8. Do the triple integral over all space

$$\iiint \psi^*\psi\, dV = \frac{1}{\pi a_0^3} \int_0^{2\pi} \int_0^{\pi} \int_0^{\infty} r^2 \sin\theta\, e^{-2r/a_0}\, dr\, d\theta\, d\phi$$

The ϕ integral yields 2π, and the θ integral yields 2. This leaves

$$\iiint \psi^*\psi\, dV = \frac{4\pi}{\pi a_0^3} \int_0^{\infty} r^2 e^{-2r/a_0}\, dr = \frac{4}{a_0^3} \frac{2}{(2/a_0)^3} = 1$$

as required.

10. $n = 3$ and $\ell = 1$, so $m_\ell = 0$ or ± 1. Thus $L_z = 0$ or $\pm\hbar$

$$L = \sqrt{\ell(\ell+1)}\hbar = \sqrt{2}\,\hbar$$

L_y and L_x are unrestricted except for the constraint $L_x^2 + L_y^2 = L^2 - L_z^2$.

15.

$$\cos\theta = \frac{L_z}{L} = \frac{m_\ell}{\sqrt{\ell(\ell+1)}}$$

For this extreme case we could have $\ell = m_\ell$ so

$$\cos(3°) = \frac{\ell}{\sqrt{\ell(\ell+1)}} \qquad \cos^2(3°) = \frac{\ell^2}{\ell(\ell+1)} = \frac{\ell^2}{\ell^2+\ell}$$

Rearranging we find

$$\ell = \left(\frac{1}{\cos^2 (3°)} - 1 \right)^{-1} = 364.1$$

and we have to round up in order to get within $3°$, so $\ell = 365$.

17.

$$\psi_{21-1} = R_{21} Y_{1-1} = \frac{1}{8\sqrt{\pi}\, a_0^{3/2}} \left(\frac{r}{a_0} \right) e^{-r/2a_0} \sin\theta e^{-i\phi}$$

$$\psi_{210} = R_{21} Y_{10} = \frac{1}{4\sqrt{2\pi}\, a_0^{3/2}} \left(\frac{r}{a_0} \right) e^{-r/2a_0} \cos\theta$$

$$\psi_{32-1} = R_{32} Y_{2-1} = \frac{1}{81\sqrt{\pi}\, a_0^{3/2}} \left(\frac{r^2}{a_0^2} \right) e^{-r/3a_0} \sin\theta \cos\theta e^{-i\phi}$$

21. Differentiating $E = \dfrac{hc}{\lambda}$ we find

$$dE = -\frac{hc}{\lambda^2} d\lambda \qquad \text{or} \qquad |\Delta E| = \frac{hc}{\lambda^2} |\Delta\lambda|.$$

In the Zeeman effect between adjacent m_ℓ states $|\Delta E| = \mu_B B$ so
$\mu_B B = \left(\dfrac{hc}{\lambda_0^2} \right) |\Delta\lambda|$ or

$$\Delta\lambda = \frac{\lambda_0^2 \mu_B B}{hc}$$

24. From Problem 21

$$\Delta\lambda = \frac{\lambda_0^2 \mu_B B}{hc}$$

so the magnetic field is

$$B = \frac{hc\, \Delta\lambda}{\lambda_0^2 \mu_B} = \frac{(1240 \text{ eV} \cdot \text{nm})\,(0.04 \text{ nm})}{(656.5 \text{ nm})^2\,(5.788 \times 10^{-5} \text{ eV/T})} = 1.99 \text{ T}$$

28. As shown in Figure 7.9 the electron spin vector cannot point in the direction of $\vec{B}$, because its magnitude is $S = \sqrt{s(s+1)} = \sqrt{3/4}\,\hbar$ and its z-component is $S_z = m_s\hbar = \hbar/2$. If the z-component of a vector is less than the vector's magnitude, the vector does not lie along the z-axis.

35. We must find the maxima and minima of the following function.

$$P(r) = r^2 |R(r)|^2$$
$$= A^2 e^{-r/a_0}\left(2 - \frac{r}{a_0}\right)^2 r^2 = A^2\left(4r^2 - \frac{4r^3}{a_0} + \frac{r^4}{a_0^2}\right)e^{-r/a_0}$$

To find the extrema set $\dfrac{dP}{dr} = 0$:

$$0 = -\frac{1}{a_0}\left(4r^2 - \frac{4r^3}{a_0} + \frac{r^4}{a_0^2}\right)e^{-r/a_0} + \left(8r - \frac{12r^2}{a_0} + \frac{4r^3}{a_0^2}\right)e^{-r/a_0}$$

$$0 = -\frac{r^3}{a_0^3} + \frac{8r^2}{a_0^2} - \frac{16r}{a_0} + 8$$

Letting $x = \dfrac{r}{a_0}$ the equation above can be factored into

$(x - 2)(x^2 - 6x + 4) = 0$. From the first factor we get $x = 2$ (or $r = 2a_0$), which from Figure 7.12 we can see is a minimum. The second parenthesis gives a quadratic equation with solutions $x = 3 \pm \sqrt{5}$, so $r = (3 \pm \sqrt{5})a_0$. These are both maxima.

44.

$$R = \frac{e^2}{4\pi\epsilon_0 mc^2} = \frac{1.44 \times 10^{-9}\ \text{eV}\cdot\text{m}}{511 \times 10^3\ \text{eV}} = 2.82 \times 10^{-15}\ \text{m}$$

From the angular momentum equation

$$v = \frac{3\hbar}{4mR} = \frac{3\hbar c}{4mc^2 R}c = \frac{3\,(197.33\ \text{eV}\cdot\text{nm})}{4\,(511 \times 10^3\ \text{eV})\,(2.82 \times 10^{-6}\ \text{nm})}c = 103c$$

A speed of $103c$ is prohibited by the rules of relativity.

49. The ground state energy can be obtained using the standard Rydberg formula with the reduced mass μ of the muonic atom

$$E_0 = \frac{e^2}{8\pi\epsilon_0 a_0} = \frac{\mu e^4}{2\left(4\pi\epsilon_0\right)^2 \hbar^2}$$

Computing the reduced mass:

$$\mu = \frac{m_p m_\mu}{m_p + m_\mu} = \frac{1}{c^2}\frac{(938.27 \text{ MeV})\,(105.66 \text{ MeV})}{(938.27 \text{ MeV} + 105.66 \text{ MeV})} = 94.966 \text{ MeV}/c^2$$

Thus

$$\begin{aligned}
E_0 &= \frac{\mu e^4}{2\left(4\pi\epsilon_0\right)^2 \hbar^2} = \left(\frac{e^2}{4\pi\epsilon_0}\right)^2 \frac{\mu c^2}{2\left(\hbar^2 c^2\right)} \\
&= \frac{(1.44 \text{ eV}\cdot\text{nm})^2\,(94.966 \times 10^6 \text{ eV})}{2\,(197.33 \text{ eV}\cdot\text{nm})^2} = 2.53 \text{ keV}
\end{aligned}$$

50. The interaction between the magnetic moment of the proton and magnetic moment of the electron causes hyperfine splitting. The transition between the two states causes emission of a photon with energy of 5.9×10^{-6} eV. From the uncertainty principle, we know $\Delta E \Delta t \geq \frac{\hbar}{2}$. With a lifetime of $\Delta t = 1 \times 10^7$ y then

$$\Delta E \geq \frac{6.5821 \times 10^{-16} \text{ eV}}{2\left(1 \times 10^7 \text{ y}\right)\left(3.16 \times 10^7 \text{ s/y}\right)} = 1.041 \times 10^{-30} \text{ eV}$$

Chapter 8

1. The first two electrons are in the $1s$ subshell and have $\ell = 0$, with $m_s = \pm 1/2$. The third electron is in the $2s$ subshell and has $\ell = 0$, with either $m_s = 1/2$ or $-1/2$. With four particles there are six possible interactions: the nucleus with electrons 1, 2, 3; electron 1 with electron 2; electron 1 with electron 3; or electron 2 with electron 3. In each case it is possible to have a Coulomb interaction and a magnetic moment interaction.

7. From Figure 8.4 we see that the radius of Na is about 0.19 nm. We know that for single-electron atoms

$$E = -\frac{Ze^2}{8\pi\varepsilon_0 r}$$

Therefore

$$Ze = -\frac{8\pi\varepsilon_0 r E}{e} = -\frac{2\,(0.19\text{ nm})\,(-5.14\text{ eV})}{1.44\text{ V} \cdot \text{nm}} = 1.36e$$

10. Pd: $[\text{Kr}]\,4d^{10}$, Hf: $[\text{Xe}]\,4f^{14}5d^26s2$, Sb: $[\text{Kr}]\,4d^{10}5s^25p^3$ where the bracket represents a closed inner shell. For example, $[\text{Kr}]$ represents $1s^22s^22p^63s^23p^64s^23d^{10}4p^6$.

15. In the $3d$ state $\ell = 2$ and $s = 1/2$, so $j = 5/2$ or $3/2$. As usual $m_\ell = 0, \pm 1, \pm 2$. The value of m_j ranges from $-j$ to j, so its possible values are $\pm 1/2$, $\pm 3/2$, and $\pm 5/2$. As always $m_s = \pm 1/2$. The two possible term notations are $3D_{5/2}$ and $3D_{3/2}$.

18. a) The quantum number m_J ranges from $-J$ to J, or $-7/2$ to $+7/2$. Then $J_z = m_J \hbar = \pm \hbar/2,\ \pm 3\hbar/2,\ \pm 5\hbar/2,\ \pm 7\hbar/2$.

   b) The minimum angle occurs when m_J is at its maximum value, which is $+7/2$. Then $J_z = 7\hbar/2$ and

$$\cos\theta = \frac{J_z}{J} = \frac{m_J \hbar}{\sqrt{J(J+1)}\,\hbar} = \frac{7/2}{\sqrt{(7/2)(9/2)}} = \frac{\sqrt{7}}{3}$$

so $\theta = 28.1°$.

21.

$$\Delta E = \frac{hc}{\lambda_1} - \frac{hc}{\lambda_2} = (1240 \text{ eV} \cdot \text{nm}) \left(\frac{1}{766.41 \text{ nm}} - \frac{1}{769.90 \text{ nm}} \right)$$
$$= 7.334 \times 10^{-3} \text{ eV}$$

As in Example 8.8 the internal magnetic field is

$$B = \frac{m\Delta E}{e\hbar} = \frac{(9.109 \times 10^{-31} \text{ kg}) (7.334 \times 10^{-3} \text{ eV})}{(1.602 \times 10^{-19} \text{ C}) (6.582 \times 10^{-16} \text{ eV} \cdot \text{s})} = 63.4 \text{ T}$$

24. As in Example 8.8

$$\Delta E = \frac{e\hbar B}{m} = \frac{(1.602 \times 10^{-19} \text{ C}) (6.582 \times 10^{-16} \text{ eV} \cdot \text{s}) (1.7 \text{ T})}{9.109 \times 10^{-31} \text{ kg}}$$
$$= 1.97 \times 10^{-4} \text{ eV}$$

26. The minimum angle corresponds to the maximum value of J_z and hence the maximum value of m_j, which is $m_j = j$. Then

$$\cos\theta = \frac{J_z}{J} = \frac{m_j\hbar}{\sqrt{j(j+1)}\,\hbar} = \frac{j}{\sqrt{j(j+1)}}$$

Solving for j we find

$$j = \frac{1}{\frac{1}{\cos^2\theta} - 1} = 2.50 = 5/2$$

33. a) $L = 0$, $S = J = 1/2$

$$g = 1 + \frac{(1/2)(3/2) + (1/2)(3/2)}{2(1/2)(3/2)} = 1 + 1 = 2$$

b) $L = 1$, $S = 1/2$, and $J = 3/2$

$$g = 1 + \frac{(3/2)(5/2) + (1/2)(3/2) - 1(2)}{2(3/2)(5/2)} = 1 + \frac{1}{3} = \frac{4}{3}$$

c) $L = 2$, $S = 1/2$, and $J = 5/2$

$$g = 1 + \frac{(5/2)(7/2) + (1/2)(3/2) - 2(3)}{2(5/2)(7/2)} = 1 + \frac{1}{5} = \frac{6}{5}$$

38. a) Magnesium has two $3s$ subshell electrons outside a closed $2p^6$ subshell. So the electronic configuration is $1s^2 2s^2 2p^6 3s^2$. Aluminum has one $3p$ electron outside the closed $3s$ level of magnesium, so its electronic configuration is $1s^2 2s^2 2p^6 3s^2 3p^1$.

b) The LS coupling for magnesium can be determined since both outer shell electrons have $\ell = 0$, then $L = 0$ and $S = 0$ since one electron will have spin up and one will have spin down. Then $J = L + S$ will also be 0. Therefore in spectroscopic notation, $n^{2S+1}L_J$, magnesium will be 3^1S_0. The $3p$ electron for aluminum will have $\ell = 1$. Therefore $L = 1$ and $S = 1/2$. Now $J = L \pm S$, so $J = 3/2$ or $J = 1/2$. The state with lower J has lower energy, so in spectroscopic notation, aluminum is $3^2P_{1/2}$.

Chapter 9

5. a)

$$\int_c^\infty F(v)\, dv = 4\pi C \int_c^\infty v^2 \exp\left(-\frac{1}{2}\beta m v^2\right) dv$$

with $T = 293$ K and $C = \left(\dfrac{\beta m}{2\pi}\right)^{3/2}$.

b) For example for H_2 gas at $T = 293$ K we have

$$\frac{1}{2}\beta m c^2 = \frac{(1)(2)\left(938 \times 10^6 \text{ eV}\right)}{2\left(8.62 \times 10^{-5} \text{ eV/K}\right)(293 \text{ K})} = 3.7 \times 10^{10}$$

The exponential of the negative of this value $\exp\left(-3.7 \times 10^{10}\right)$ is almost zero.

7. a)

$$\bar{v} = \frac{4}{\sqrt{2\pi}}\sqrt{\frac{kT}{m}} = \frac{4}{\sqrt{2\pi}}\sqrt{\frac{\left(1.381 \times 10^{-23} \text{ J/K}\right)(300 \text{ K})}{\left(1.675 \times 10^{-27} \text{ kg}\right)}} = 2510 \text{ m/s}$$

$$v^* = \sqrt{\frac{2kT}{m}} = \sqrt{\frac{2\left(1.381 \times 10^{-23} \text{ J/K}\right)(300 \text{ K})}{\left(1.675 \times 10^{-27} \text{ kg}\right)}} = 2220 \text{ m/s}$$

b)

$$\bar{v} = \frac{4}{\sqrt{2\pi}}\sqrt{\frac{kT}{m}} = \frac{4}{\sqrt{2\pi}}\sqrt{\frac{\left(1.381 \times 10^{-23} \text{ J/K}\right)(2000 \text{ K})}{\left(1.675 \times 10^{-27} \text{ kg}\right)}} = 6480 \text{ m/s}$$

$$v^* = \sqrt{\frac{2kT}{m}} = \sqrt{\frac{2\left(1.381 \times 10^{-23} \text{ J/K}\right)(2000 \text{ K})}{\left(1.675 \times 10^{-27} \text{ kg}\right)}} = 5740 \text{ m/s}$$

14. a) Use Equation (9.8) for the translational kinetic energy of one atom and multiply by Avagadro's number for one mole.

$$K = N_A \left(\frac{3}{2}kT\right) = \left(6.022 \times 10^{23}\right)\left(\frac{3}{2}\left(1.381 \times 10^{-23} \text{ J/K}\right)(273 \text{ K})\right)$$

$$= 3406 \text{ J}$$

b) Since the translational kinetic energy depends only on temperature and one mole of anything contains the same number of objects, the answer is the same for argon or oxygen.

16. Starting with the distribution

$$F(E) = \frac{8\pi C}{\sqrt{2}m^{3/2}} E^{1/2} \exp(-\beta E)$$

and setting $\dfrac{dF}{dE} = 0$, we get

$$0 = \frac{d}{dE}\left[E^{1/2}\exp(-\beta E)\right] = \frac{1}{2}E^{-1/2}\exp(-\beta E) - \beta E^{1/2}\exp(-\beta E)$$

Thus $0 = E^{-1/2} - 2\beta E^{1/2}$ which solving for E gives the desired $E^* = \dfrac{kT}{2}$.

17. The ratio of the numbers on the two levels is

$$\frac{n_2(E)}{n_1(E)} = \frac{8\exp(-\beta E_2)}{2\exp(-\beta E_1)} = 4\exp(-\beta(E_2 - E_1)) = 10^{-6}$$

$$\exp(-\beta(E_2 - E_1)) = 2.5 \times 10^{-7}$$

Taking logarithms:

$$-\beta(E_2 - E_1) = -\frac{E_2 - E_1}{kT} = \ln(2.5 \times 10^{-7}) = -15.20$$

For atomic hydrogen $E_2 - E_1 = \frac{3}{4}E_0 = 10.20$ eV. Finally

$$T = -\frac{E_2 - E_1}{k(-15.20)} = -\frac{10.20 \text{ eV}}{(8.618 \times 10^{-5} \text{ eV/K})(-15.20)} = 7790 \text{ K}$$

22. At first one may think it should be 0.5, but this is not quite true, due to the asymmetric shape of the distribution. Starting with

Equation (9.43) for $g(E)$ and using the fact that $F_{\text{FD}} \approx 1$ in this range, we have

$$
N\,(E < E_{\text{F}}) \;=\; \int_0^E g(E)(1)\,dE = \frac{3}{2} N E_{\text{F}}^{-3/2} \int_0^E E^{1/2}\,dE
$$
$$
= \; N E_{\text{F}}^{-3/2} \overline{E}^{3/2}
$$

But recalling that $\overline{E} = \frac{3}{5} E_{\text{F}}$, we see that

$$
N\,(E < E_{\text{F}}) = N \left(\frac{3}{5}\right)^{3/2} = 0.465\,N
$$

25. a) As in Problem 23, $N/V = 5.86 \times 10^{28}$ m^{-3}. Then

$$
E_{\text{F}} \;=\; \frac{h^2}{8m} \left(\frac{3}{\pi}\frac{N}{V}\right)^{2/3}
$$
$$
= \; \frac{(6.626 \times 10^{-34}\ \text{J}\cdot\text{s})^2}{8\,(9.109 \times 10^{-31}\ \text{kg})} \left(\frac{3}{\pi}\,(5.86 \times 10^{28}\ \text{m}^{-3})\right)^{2/3}
$$
$$
= \; 8.81 \times 10^{-19}\ \text{J} \; = 5.50\ \text{eV}
$$

b)

$$
u_{\text{F}} = \sqrt{\frac{2E_{\text{F}}}{m}} = \sqrt{\frac{2\,(8.81 \times 10^{-19}\ \text{J})}{9.109 \times 10^{-31}\ \text{kg}}} = 1.39 \times 10^6\ \text{m/s}
$$

29. In general $E_{\text{F}} = \frac{1}{2}mu_{\text{F}}^2$, so $u_{\text{F}} = \sqrt{2E_{\text{F}}/m}$.

a)

$$
u_{\text{F}} \;=\; \sqrt{\frac{2E_{\text{F}}}{m}} = \sqrt{\frac{2\,(3.93\ \text{eV})\,(1.602 \times 10^{-19}\ \text{J/eV})}{9.109 \times 10^{-31}\ \text{kg}}}
$$
$$
= \; 1.18 \times 10^6\ \text{m/s}
$$

b)

$$
u_{\text{F}} \;=\; \sqrt{\frac{2E_{\text{F}}}{m}} = \sqrt{\frac{2\,(9.47\ \text{eV})\,(1.602 \times 10^{-19}\ \text{J/eV})}{9.109 \times 10^{-31}\ \text{kg}}}
$$
$$
= \; 1.83 \times 10^6\ \text{m/s}
$$

37. We assume that the collection of fermions behaves like an ideal gas. Using Maxwell-Boltzmann statistics, we know that $E = \frac{3}{2}kT$ or $\beta E = 3/2$. We note that $\exp(\beta E) = \exp(3/2) = 4.4817$. Now the MB factor is $\exp(-\beta E)$ and we want this to be within 1% of the FD factor: $F_{FD} = \dfrac{1}{\exp\left[\frac{(E-E_F)}{kT}\right]+1}$. So we want

$$\exp[\beta(E-E_F)]+1 = 4.5265 \quad \text{or} \quad \beta(E-E_F) = \ln(3.5265).$$

At room temperature $(\beta)^{-1} = 2.525 \times 10^{-2}$ eV so the expression above can be solved to give $E - E_F = 3.2 \times 10^{-2}$ eV. This small value is possible at room temperature.

38. a) To find N/V integrate $n(E)\,dE$ over the whole range of energies:

$$\frac{N}{V} = \left(\frac{1}{L^3}\right)\int_0^\infty n(E)\,dE = \frac{8\pi}{h^3c^3}\int_0^\infty \frac{E^2}{\exp(E/kT)-1}\,dE$$

From integral tables we have the following:

$$\int_0^\infty \frac{x^{n-1}}{e^{mx}-1}\,dx = m^{-n}\Gamma(n)\zeta(n)$$

For us $m = \dfrac{1}{kT}$, $\Gamma(3) = 2! = 2$, and from numerical tables $\zeta(3) \approx 1.20$. Thus

$$\frac{N}{V} = \frac{8\pi}{h^3c^3}(kT)^3(2)(1.20) = \frac{8\pi k^3 T^3}{h^3c^3}(2.40)$$

b) With $T = 500$ K:

$$\begin{aligned}
\frac{N}{V} &= \frac{8\pi k^3 T^3}{h^3c^3}(2.40) \\
&= 8\pi(2.40)\left(\frac{(1.381 \times 10^{-23}\text{ J/K})(500\text{ K})}{(6.626 \times 10^{-34}\text{ J}\cdot\text{s})(2.998 \times 10^8\text{ m/s})}\right)^3 \\
&= 2.53 \times 10^{15}\text{ m}^{-3}
\end{aligned}$$

At $T = 5500$ K:

$$\frac{N}{V} = \frac{8\pi k^3 T^3}{h^3 c^3}(2.40)$$

$$= 8\pi(2.40)\left(\frac{(1.381 \times 10^{-23} \text{ J/K})(5500 \text{ K})}{(6.626 \times 10^{-34} \text{ J}\cdot\text{s})(2.998 \times 10^8 \text{ m/s})}\right)^3$$

$$= 3.37 \times 10^{18} \text{ m}^{-3}$$

45. The number of molecules with speed v that hit the wall per unit time is proportional to v and $F(v)$, so that the distribution $W(v)$ of the escaping molecules is by proportion

$$W(v) \sim vF(v) \sim v^3 \exp\left(-\frac{1}{2}\beta mv^2\right)$$

Let the normalization constant for $W(v)$ be C', so

$$C' \int_0^\infty v^3 \exp\left(-\frac{1}{2}\beta mv^2\right) dv = 1 = C'\left(\frac{1}{2}\right)\left(\frac{\beta m}{2}\right)^{-2}$$

or $C' = \beta^2 m^2/2$. The mean kinetic energy of the escaping molecules is

$$\overline{E} = \frac{1}{2}m\overline{v^2} = \frac{1}{2}mC' \int_0^\infty v^5 \exp\left(-\frac{1}{2}\beta mv^2\right) dv$$

$$= \frac{1}{2}m\left(\frac{\beta^2 m^2}{2}\right)\left(\frac{\beta m}{2}\right)^{-3} = \frac{2}{\beta} = 2kT$$

47. For the harmonic oscillator the position and velocity are

$$x = x_0 \cos(\omega t) \qquad\qquad v = \frac{dx}{dt} = -\omega x_0 \sin(\omega t)$$

$$V = \frac{1}{2}kx^2 = \frac{1}{2}kx_0^2 \cos^2(\omega t)$$

$$K = \frac{1}{2}mv^2 = \frac{1}{2}m\omega^2 x_0^2 \sin^2(\omega t) = \frac{1}{2}kx_0^2 \sin^2(\omega t)$$

where we have used the fact that $\omega^2 m = k$. Over one cycle the average of the square of the sine or cosine function is one-half. Also the total energy is $E = \frac{1}{2}kx_0^2$. Thus

$$\overline{K} = \overline{V} = \frac{1}{2}kx_0^2\left(\frac{1}{2}\right) = \frac{E}{2}$$

49. Rearranging equation (9.64) and with $m = 4\,\text{u}$ we have

$$\frac{N}{V} \leq \frac{2\pi\,(2.315)}{h^3}\,[2mkT]^{3/2}$$

The term in brackets equals

$$\left[2\,(4)\,(1.6605 \times 10^{-27}\,\text{kg})\,(1.381 \times 10^{-23}\,\text{J/K})\,(293\,\text{K})\right]^{3/2}$$

$$\doteq 3.94 \times 10^{-70}\,(\text{J/kg})^{3/2}$$

so the expression equals

$$\leq \frac{2\pi\,(2.315)}{(6.626 \times 10^{-34}\,\text{J}\cdot\text{s})^3}\left[3.94 \times 10^{-70}\,(\text{J/kg})^{3/2}\right]$$

$$\leq 1.97 \times 10^{31}\,\text{m}^{-3}$$

The number density of an ideal gas at STP is $\dfrac{N}{V} = \dfrac{P}{kT}$

$$= \frac{1.0135 \times 10^5\,\text{Pa}}{(1.381 \times 10^{-23}\,\text{J/K})\,(293\,\text{K})}. \text{ Therefore } \frac{N}{V} = 2.50 \times 10^{25}\,\text{m}^{-3}. \text{ As}$$

you would expect, the condensate has a number density nearly one million times greater than the ideal gas.

Chapter 10

1. a) For each state the energy is given by $E_{\text{rot}} = \dfrac{\hbar^2 \ell(\ell+1)}{2I}$, so the transition energy is

$$\Delta E = \frac{\hbar^2}{2I}(2(3) - 1(2)) = \frac{2\hbar^2}{I} = \frac{2\left(1.055 \times 10^{-34} \text{ J} \cdot \text{s}\right)^2}{10^{-46} \text{ kg} \cdot \text{m}^2}$$
$$= 2.2 \times 10^{-22} \text{ J} = 1.4 \times 10^{-3} \text{ eV}$$

b) As in part (a)

$$\Delta E = \frac{\hbar^2}{2I}(15(16) - 14(15)) = \frac{15\hbar^2}{I} = 1.67 \times 10^{-21} \text{ J}$$
$$= 1.04 \times 10^{-2} \text{ eV}$$

This is still in the infrared part of the spectrum.

5. With Bohr's condition $L = n\hbar$ we find

$$E_{\text{rot}} = \frac{L^2}{2I} = \frac{n^2\hbar^2}{2I}$$

The Bohr version and the correct version become similar for large values of quantum number n or ℓ, but they are quite different for small ℓ.

6. $\Delta E = E_1 - E_0 = E_1 = \dfrac{\hbar^2}{I} = \dfrac{hc}{\lambda}.$

$$I = \frac{\hbar^2 \lambda}{hc} = \frac{\hbar\lambda}{2\pi c} = \frac{\left(1.055 \times 10^{-34} \text{ J} \cdot \text{s}\right)\left(1.3 \times 10^{-3} \text{ m}\right)}{2\pi\left(2.998 \times 10^8 \text{ m/s}\right)}$$
$$= 7.28 \times 10^{-47} \text{ kg} \cdot \text{m}^2$$

b) The minimum energy in a vibrational transition is $\Delta E = hf$. From Table 10.1 $f = 6.42 \times 10^{13}$ Hz, which corresponds to a photon of wavelength $\lambda = c/f = 4.67\mu$m. A photon of this wavelength or less is required to excite the vibrational mode, so the 1.30 mm photon is too weak.

10. a) The distance of each H atom from the line is
$d = (0.0958 \text{ nm}) (\sin 52.5°) = 7.60 \times 10^{-2}$ nm. Then

$$
\begin{aligned}
I &= 2m_H d^2 = 2 \left(1.67 \times 10^{-27} \text{ kg}\right) \left(7.60 \times 10^{-11} \text{ m}\right)^2 \\
&= 1.93 \times 10^{-47} \text{ kg} \cdot \text{m}^2
\end{aligned}
$$

b)

$$
E_1 = \frac{\hbar^2}{I} = \frac{\left(1.055 \times 10^{-34} \text{ J} \cdot \text{s}\right)^2}{1.93 \times 10^{-47} \text{ kg} \cdot \text{m}^2} = 5.77 \times 10^{-22} \text{ J} = 3.61 \text{ MeV}
$$

$$
E_2 = \frac{3\hbar^2}{I} = \frac{3 \left(1.055 \times 10^{-34} \text{ J} \cdot \text{s}\right)^2}{1.93 \times 10^{-47} \text{ kg} \cdot \text{m}^2} = 1.73 \times 10^{-21} \text{ J} = 10.81 \text{ MeV}
$$

c)

$$
\lambda = \frac{hc}{E_1} = \frac{\left(6.626 \times 10^{-34} \text{ J} \cdot \text{s}\right) \left(2.998 \times 10^8 \text{ m/s}\right)}{5.77 \times 10^{-22} \text{ J}} = 344 \, \mu\text{m}
$$

15. The gap between adjacent lines is $h (\Delta f) = \hbar^2/I$.

a)

$$
I = \frac{\hbar^2}{h (\Delta f)} = \frac{\left(1.055 \times 10^{-34} \text{ J} \cdot \text{s}\right)^2}{\left(6.626 \times 10^{-34} \text{ J} \cdot \text{s}\right) \left(7 \times 10^{11} \text{ Hz}\right)} = 2.4 \times 10^{-47} \text{ kg} \cdot \text{m}^2
$$

b) $\omega = 2\pi f = \sqrt{\kappa/\mu}$ with $f = 8.65 \times 10^{13}$ Hz. The reduced mass is (using the ^{35}Cl isotope)

$$
\mu = \frac{m_1 m_2}{m_1 + m_2} = \frac{35}{36} u = 1.614 \times 10^{-27} \text{ kg}
$$

Solving for κ we find

$$
\kappa = 4\pi^2 f^2 \mu = 4\pi^2 \left(8.65 \times 10^{13} \text{ Hz}\right)^2 \left(1.614 \times 10^{-27} \text{ kg}\right) = 478 \text{ N/m}
$$

in good agreement with Table 10.1.

18. a) Using dimensional analysis and the fact that the energy of each photon is $hc/\lambda = 3.14 \times 10^{-19}$ J,

$$\frac{N}{t} = \frac{5 \times 10^{-3} \text{ J/s}}{3.14 \times 10^{-19} \text{ J/photon}} = 1.59 \times 10^{16} \text{ photon/s}$$

b) 0.02 mole is equal to $0.02 N_A = 1.20 \times 10^{22}$ atoms. Then the fraction participating is

$$\frac{1.59 \times 10^{16}}{1.20 \times 10^{22}} = 1.33 \times 10^{-6}$$

c) The transitions involved have a fairly low probability, even with stimulated emission. We are saved by the large number of atoms available.

25. Using dimensional analysis the number density is

$$1980 \text{ kg/m}^3 \frac{1 \text{ mol}}{0.07455 \text{ kg}} \frac{2 \left(6.022 \times 10^{23}\right)}{\text{mol}} = 3.20 \times 10^{28} \text{ m}^{-3}$$

Therefore the distance is

$$d = \left(3.20 \times 10^{28} \text{ m}^{-3}\right)^{-1/3} = 3.15 \times 10^{-10} \text{ m} = 0.315 \text{ nm}$$

26. Each charge has two unlike charges a distance r away, two like charges a distance $2r$ away, and so on:

$$V = -\frac{2e^2}{4\pi\epsilon_0 r} \left(1 - \frac{1}{2} + \frac{1}{3} - \frac{1}{4} + ...\right)$$

The bracketed expression is the Taylor series expansion for $\ln 2$, so

$$V = -\frac{2e^2}{4\pi\epsilon_0 r} \ln 2 = -\frac{\alpha e^2}{4\pi\epsilon_0 r}$$

and we see that $\alpha = 2 \ln 2$.

32. We begin with Equation (10.32); $L \equiv \dfrac{K}{\sigma T} = \dfrac{4k^2}{\pi\,e^2}$. Solving for the thermal conductivity, K, we find

$$
\begin{aligned}
K &= \frac{4\,\sigma k^2 T}{\pi\,e^2} \\[2mm]
&= \frac{4\left(6.0 \times 10^7\,\Omega^{-1}\cdot\mathrm{m}^{-1}\right)\left(1.3807 \times 10^{-23}\,\mathrm{J}\cdot\mathrm{K}^{-1}\right)^2 293\,\mathrm{K}}{\pi\left(1.6022 \times 10^{-19}\,\mathrm{C}\right)^2} \\[2mm]
&= 166\,\mathrm{W}\cdot\mathrm{K}^{-1}\cdot\mathrm{m}^{-1}
\end{aligned}
$$

As mentioned in the text, the Wiedemann-Franz Law was derived using classical expressions for the mean speed and the molar heat capacity. Quantum mechanical corrections give an additional factor of $\dfrac{\pi^3}{12} = 2.58$. With this correction, our answer would be $428\,\mathrm{W}\cdot\mathrm{K}^{-1}\cdot\mathrm{m}^{-1}$ which is much closer to the measured value.

39. a) Following the arguments in the text, regardless of the sense of rotation of the electron, as the magnetic field increases from zero, the flux upward through the loop increases. Then, as in the text, the tangential electric field is still directed clockwise. The torque does not depend on the sense of rotation either, so the torque is directed out of the page. Thus the ΔL is out of the page which means that $\Delta\mu = -\dfrac{e}{2m}\Delta L = \dfrac{e^2 r^2 B}{4m}$ is *into* the page and thus opposite the direction of B as before. (If the reader is uncertain that the sense of rotation is irrelevant, recall that $\Phi_B = \int \vec{B}\cdot d\vec{a}$. You use the right-hand rule to determine the sense of $d\vec{a}$. While the direction of $d\vec{a}$ will depend on the sense of rotation, the other side of the equation for Faraday's law contains the expression $\oint \vec{E}\cdot \vec{d\ell}$. The sense of transit along the increment $\vec{d\ell}$ will also change.)

b) With the field directed into the paper, now the magnetic flux increases downward through the loop. Thus Faraday's law will indicate that the electric field is tangent to the orbit but now directed counterclockwise. Thus the torque will directed into the page and thus ΔL will be directed into the page. Finally the negative sign in $\Delta\mu = -\dfrac{e}{2m}\Delta L$ will mean that $\Delta\mu = \dfrac{e^2 r^2 B}{4m}$ is directed *out of* the page.

43. The magnetic dipole moment has units A·m^2, so M has units A·m^2/m^3 = A/m. μ_0 has units T·m/A and B has units T, so $\chi = \dfrac{\mu_0 M}{B}$ has units $\dfrac{(\text{T} \cdot \text{m/A})(\text{A/m})}{\text{T}}$ which reduces to no units.

45.

$$B_c = B_c(0)\left(1 - \left(\frac{T}{T_c}\right)^2\right) = 0.1 B_c(0)$$

Thus $(T/T_c)^2 = 0.9$ and $T = \sqrt{0.9}T_c \approx 0.95 T_c$.

Similarly for a ratio of 0.5 we find $T = \sqrt{0.5}T_c \approx 0.71 T_c$, and for a ratio of 0.9 we find $T = \sqrt{0.1}T_c \approx 0.32 T_c$.

47. Using the value given in the text just below Equation (10.43), $T_c = 4.146$ K for a mass of 203.4 u, we find

$$M^{0.5}T_c = \text{constant} = 59.1296 \; \text{u}^{0.5} \cdot \text{K}$$

For a mass of 201 u we find $T_c = 4.171$ K and for a mass of 204 u we find $T_c = 4.140$ K.

50.

$$B = \mu_0 I n = \left(4\pi \times 10^{-7} \; \text{N/A}^2\right)(5.0 \; \text{A})(2500 \; \text{m}^{-1}) = 15.71 \, \text{mT}$$

$$\Phi = BA = \frac{B\pi d^2}{4} = \left(15.71 \times 10^{-3} \; \text{T}\right)\frac{\pi (0.028 \; \text{m})^2}{4}$$

$$= 9.67 \times 10^{-6} \; \text{T} \cdot \text{m}^2$$

$$\frac{\Phi}{\Phi_0} = \frac{9.67 \times 10^{-6} \; \text{T} \cdot \text{m}^2}{2.068 \times 10^{-15} \; \text{T} \cdot \text{m}^2} = 4.7 \times 10^9 \; \text{flux quanta}$$

This large number shows how small the flux quantum is.

58. a) In a RL circuit the current is

$$I = I_0 e^{-Rt/L}$$

For small values of R let us approximate the exponential with the Taylor expansion $1 - Rt/L$. Then

$$10^{-9} = 1 - \frac{I}{I_0} = 1 - e^{-Rt/L} \approx \frac{Rt}{L}$$

$$R \leq 10^{-9} \frac{L}{t} = 10^{-9} \left(\frac{3.14 \times 10^{-8} \text{ H}}{2.5 \text{ y } (3.16 \times 10^7 \text{ s/y})} \right) = 4.0 \times 10^{-25} \text{ } \Omega$$

b) For a 10% loss

$$t = \frac{0.1}{10^{-9}} (2.5 \text{ y}) = 2.5 \times 10^8 \text{ y}$$

Chapter 11

4. a) Starting from Equation (11.6) and with $A = yz$, we have

$$n = \frac{IB}{eV_H z} = \frac{(0.10 \text{ A})(0.036 \text{ T})}{(1.602 \times 10^{-19} \text{ C})(8.4 \times 10^{-3} \text{ V})(1.5 \times 10^{-4} \text{ m})}$$
$$= 1.78 \times 10^{22} \text{ m}^{-3}$$

b) Graphing B vs. V_H we find a slope of approximately 4.56 T/V. Algebraically, we see that $B = \frac{nez}{I} V_H$. Thus the slope, m is equal to $\frac{nez}{I}$. So

$$n = \frac{mI}{ez} = \frac{(4.56 \text{ T/V})(0.10 \text{ A})}{(1.602 \times 10^{-19} \text{ C})(1.5 \times 10^{-4} \text{ m})} = 1.90 \times 10^{22} \text{ m}^{-3}$$

8. a) Al is in Group III; Ge is in Group IV; so relative to Ge, it is p-type.

b) Se has two more outer electrons than Si, so it is n-type. Se is in Group VI; Ge is in Group IV.

13. In general $I = I_0 \left(\exp(eV/kT) - 1 \right)$ and in Example 11.4

$$I_0 = \frac{I}{\exp(eV/kT) - 1} \approx 18.16 \, \mu\text{A}$$

a)

$$I = \left(18.16 \times 10^{-6} \text{A} \right) \left(\exp \left(\frac{0.250 \text{ eV}}{(8.617 \times 10^{-5} \text{ eV/K})(250 \text{ K})} \right) - 1 \right)$$
$$= 1.99 \text{ A}$$

b)

$$I = \left(18.16 \times 10^{-6} \text{A} \right) \left(\exp \left(\frac{0.250 \text{ eV}}{(8.617 \times 10^{-5} \text{ eV/K})(300 \text{ K})} \right) - 1 \right)$$
$$= 0.288 \text{ A} = 288 \text{ mA}$$

c)

$$I = \left(18.16 \times 10^{-6}\text{A}\right)\left(\exp\left(\frac{0.250 \text{ eV}}{(8.617 \times 10^{-5} \text{ eV/K})(500 \text{ K})}\right) - 1\right)$$

$$= 0.006 \text{ A} = 6.00 \text{ mA}$$

14. Since the tube is single-walled, we can find the surface area density, σ in SI units. Each atom has mass 12 u, with $u = 1.6605 \times 10^{-27}$ kg.

$$\sigma = \left(2.3 \times 10^{19} \text{ atoms/m}^2\right)(12 \text{ u})\left(1.6605 \times 10^{-27} \text{ kg/u}\right)$$

$$= 4.58 \times 10^{-7} \text{ kg/m}^2$$

a) To determine the density of the material, we need the mass per unit volume. The mass will equal the mass density, σ from above, times the area A of the cylindrical wall. If the tube's length is L with radius R, then $m = 2\pi RL\sigma$. Therefore the density will be

$$\rho = \frac{m}{V} = \frac{2\pi RL\sigma}{\pi R^2 L} = \frac{2\sigma}{R} = \frac{2\left(4.58 \times 10^{-7} \text{ kg/m}^2\right)}{0.7 \times 10^{-9} \text{ m}} = 1300 \text{ kg/m}^3$$

b) The material is less dense than steel and has a greater tensile strength.

17. a) Replacing ϵ_0 with $\kappa\epsilon_0$ we have for the new Bohr radius

$$a_0' = \frac{4\pi\epsilon_0\kappa\hbar^2}{me^2} = 11.7a_0 = 11.7\left(5.29 \times 10^{-2} \text{ nm}\right) = 0.619 \text{ nm}$$

b) This value is about 2.6 times the lattice spacing. This is consistent with the fact that the electron is very weakly bound, and hence the doped silicon should have a higher electrical conductivity than pure silicon.

20. a) $I = I_0\left(\exp\left(eV/kT\right) - 1\right)$. To find the value of V for the diode, use the loop rule:

$V + IR = 6\,\text{V}$, so $V = 6\text{ V} - IR$. We are given that $I_0 = 1.75\,\mu\text{A}$ and $I = 80$ mA with $T = 293$ K.

$$\frac{I}{I_0} = 45714 = \exp(eV/kT)$$

$$\ln 45714 = \frac{eV}{kT} = \frac{e\,(6\text{ V} - IR)}{kT}$$

Solving for R we find

$$R = \frac{6\text{ V} - (kT\ln 45714)/e}{I}$$

$$= \frac{6\text{ V} - (8.617 \times 10^{-5}\,\text{eV/K})\,(293\,\text{K}\,(\ln 45714))/e}{0.080\,\text{A}} = 71.6\,\Omega$$

b)

$$V_R = IR = (0.080\text{ A})\,(71.6\,\Omega) = 5.73\text{ V}$$

23.

$$E_g = \frac{hc}{\lambda} = \frac{1240\text{ eV}\cdot\text{nm}}{460\text{ nm}} = 2.70\text{ eV}$$

24. a) The total area is $(12 \times 10^6)\,(1.3 \times 10^{-7}\text{ m})^2 = 2.03 \times 10^{-7}\text{ m}^2$ so each side is the square root of this or 0.450 mm.

b) The area of each transistor is now $2.5 \times 10^{-15}\text{ m}^2$, so the number is

$$N = \frac{2.03 \times 10^{-7}\text{ m}^2}{2.5 \times 10^{-15}\text{ m}^2} = 8.12 \times 10^7$$

or nearly a factor of 7 improvement.

31. The number of bits stored is $(4.7 \times 10^9)(8) = 3.76 \times 10^{10}$ bits. To find the area we use $A = \pi\,(r_2^2 - r_1^2) = \pi\,(0.058^2 - 0.023^2) = 8.91 \times 10^{-3}\text{ m}^2$. The number of bits stored per square meter is then

$$\frac{3.76 \times 10^{10}\text{ bits}}{8.91 \times 10^{-3}\text{ m}^2} = 4.2 \times 10^{12}\text{ bits/m}^2\,.$$

Chapter 12

6. ^{14}N(99.63%), ^{15}N(0.37%); ^{50}V(0.250%), ^{51}V(99.75%); ^{85}Rb(72.17%), ^{87}Rb(27.83%)

9.

$$\frac{\mu_p}{\mu_e} = \frac{2.79\mu_N}{-1.00116\mu_B} = -2.786\frac{\mu_N}{\mu_B} = -2.787\frac{m_e}{m_p}$$

$$= -2.786\frac{0.51100}{938.27} = -1.52 \times 10^{-3}$$

10. From Appendix 8 the mass of the nuclide is 55.935 u or 9.29×10^{-26} kg.

$$\rho = \frac{m}{\frac{4}{3}\pi r^3} = \frac{m}{\frac{4}{3}\pi r_0^3 A} = \frac{9.29 \times 10^{-26}\ \text{kg}}{\frac{4}{3}\pi\,(1.2 \times 10^{-15}\ \text{m})^3\,(56)} = 2.29 \times 10^{17}\ \text{kg/m}^3$$

13. The distance equals the nuclear radius:

$$r = r_0 A^{1/3} = (1.2\ \text{fm})\left(3^{1/3}\right) = 1.73\ \text{fm}$$

$$|F_g| = \frac{GMm}{r^2} = \frac{\left(6.673 \times 10^{-11}\ \text{N} \cdot \text{m}^2/\text{kg}^2\right)\left(1.673 \times 10^{-27}\ \text{kg}\right)^2}{\left(1.73 \times 10^{-15}\ \text{m}\right)^2}$$

$$= 6.24 \times 10^{-35}\ \text{N}$$

$$|F_e| = \frac{ke^2}{r^2} = \frac{\left(8.988 \times 10^9\ \text{N} \cdot \text{m}^2/\text{C}^2\right)\left(1.602 \times 10^{-19}\ \text{C}\right)^2}{\left(1.73 \times 10^{-15}\ \text{m}\right)^2} = 77.1\ \text{N}$$

To compare with the strong force, we need the potential energy:

$$|V_g| = \frac{GMm}{r} = \frac{\left(6.673 \times 10^{-11}\ \text{N} \cdot \text{m}^2/\text{kg}^2\right)\left(1.673 \times 10^{-27}\ \text{kg}\right)^2}{\left(1.73 \times 10^{-15}\ \text{m}\right)\left(1.602 \times 10^{-13}\ \text{J/MeV}\right)}$$

$$= 6.7 \times 10^{-37}\ \text{MeV}$$

$$|V_e| = \frac{ke^2}{r} = \frac{\left(8.988 \times 10^9\ \text{N} \cdot \text{m}^2/\text{C}^2\right)\left(1.602 \times 10^{-19}\ \text{C}\right)^2}{\left(1.73 \times 10^{-15}\ \text{m}\right)\left(1.602 \times 10^{-13}\ \text{J/MeV}\right)} = 0.83\ \text{MeV}$$

The electrostatic force is about 50 times weaker than the strong force. The gravitational force is almost 10^{38} times weaker than the strong force.

20. For ^{4}He the radius is

$$r = r_0 A^{1/3} = (1.2 \text{ fm}) \left(4^{1/3}\right) = 1.90 \text{ fm}$$

$$V = \frac{e^2}{4\pi\epsilon_0 r} = \frac{1.44 \times 10^{-9} \text{ eV} \cdot \text{m}}{1.90 \times 10^{-15} \text{ m}} = 0.76 \text{ MeV}$$

For ^{40}Ca:

$$\Delta E_{\text{Coul}} = \frac{3}{5}\frac{Z(Z-1)e^2}{4\pi\epsilon_0 R} = 0.72\, Z(Z-1)\, A^{-1/3} \text{ MeV}$$

$$= 0.72\,(20)\,(19)\,40^{-1/3} \text{ MeV} = 80\,\text{MeV}$$

For ^{208}Pb:

$$\Delta E_{\text{Coul}} = 0.72\,(82)\,(81)\,208^{-1/3} \text{ MeV} = 807\,\text{MeV}$$

There is a factor of ten between each of these three nuclides.

23. We begin with Equation (12.20) and substitute Equation (12.19) for the Coulomb term. Therefore, we have

$$B\left(_Z^A X\right) = a_V A - a_A A^{2/3} - 0.72\,[Z(Z-1)]\,A^{-1/3} - a_S \frac{(N-Z)^2}{A} + \delta$$

Evaluating this expression for ^{48}Ca, we have:

$$B\left(_{20}^{48}\text{Ca}\right) =$$

$$(14\,\text{MeV})\,48 - (13\,\text{MeV})\,48^{2/3} - 0.72\frac{[20 \cdot 19]}{48^{1/3}} - (19\,\text{MeV})\frac{(28-20)^2}{48}$$
$$+ \frac{33\,\text{MeV}}{48^{3/4}}$$

This gives $B\left(_{20}^{48}\text{Ca}\right) = 401.492\,\text{MeV}$ which equals

57

$401.492 \, \text{MeV} \left(\dfrac{c^2 \cdot u}{931.5 \, \text{MeV}} \right)$ which is $= 0.43102 \, u \cdot c^2$. From Equation (12.10), we can calculate the mass.

$$M \left({}^{48}_{20}\text{Ca} \right) = 28 \cdot m_n + 20 \cdot M \left({}^{1}\text{H} \right) - \left[B \left({}^{48}_{20}\text{Ca} \right) / c^2 \right]$$

Substituting, we find

$$
\begin{aligned}
M \left({}^{48}_{20}\text{Ca} \right) &= 28 \cdot (1.008665 \, u) + 20 \cdot (1.007825 \, u) - \left[0.43102 \, u \cdot c^2 / c^2 \right] \\
&= 47.9681 \, u
\end{aligned}
$$

The value equals $47.9681 \, u \cdot \left(\dfrac{931.5 \, \text{MeV}}{c^2 \cdot u} \right) = 4.468 \times 10^4 \, \text{MeV}/c^2$.
The atomic mass given in Appendix 8 is $47.952534 \, u$. Our calculation based on a semi-empirical formula is different by about 0.03%.

25.

$$\lambda = \frac{\ln 2}{t_{1/2}} = \frac{\ln 2}{(5.271 \, \text{y}) \, (3.156 \times 10^7 \, \text{s/y})} = 4.167 \times 10^{-9} \, \text{s}^{-1}$$

$$N = \frac{R}{\lambda} = \frac{2.4 \times 10^7 \, \text{s}^{-1}}{4.167 \times 10^{-9} \, \text{s}^{-1}} = 5.76 \times 10^{15}$$

$$m = \left(5.76 \times 10^{15} \right) \frac{1 \, \text{mol}}{6.022 \times 10^{23} \, \text{mol}} \frac{60 \, \text{g}}{} = 0.57 \, \mu\text{g}$$

27. In general (using the definition of the mean value of a function)

$$\tau = \frac{\int_0^\infty t \, R(t) \, dt}{\int_0^\infty R(t) \, dt} = \frac{1}{N_0} \int_0^\infty t \, R(t) \, dt$$

because all nuclei must decay between $t = 0$ and $t = \infty$. Using $R = R_0 e^{-\lambda t}$ we have

$$\tau = \frac{R_0}{N_0} \int_0^\infty t \, e^{-\lambda t} \, dt = \frac{R_0}{N_0} \frac{1}{\lambda^2} = \frac{\lambda N_0}{\lambda^2 N_0} = \frac{1}{\lambda} = \frac{t_{1/2}}{\ln 2}$$

30.

$$\lambda = \frac{\ln 2}{t_{1/2}} = \frac{\ln 2}{(109.8 \, \text{min}) \, (60 \, \text{s/min})} = 1.052 \times 10^{-4} \, \text{s}^{-1}$$

$$
\begin{aligned}
R &= R_0 e^{-\lambda t} = \left(10^7 \, \text{Bq} \right) \exp \left(- \left(1.052 \times 10^{-4} \, \text{s}^{-1} \right) (48) \, (3600 \, \text{s}) \right) \\
&= 0.127 \, \text{Bq}
\end{aligned}
$$

Chapter 12

39. ^{80}Br $\rightarrow$ ^{76}As + ^{4}He:

$Q = [79.918530 - 75.922394 - 4.002603]\,u \cdot c^2 = -6.0$ MeV

(not allowed)

^{80}Br $\rightarrow$ ^{80}Kr + β^-:

$Q = (79.918530 - 79.916378)\,u \cdot c^2 = 2.0$ MeV (allowed)

^{80}Br $\rightarrow$ ^{80}Se + β^+:

$Q = (79.918530 - 79.916522 - 2\,(0.000549))\,u \cdot c^2 = 0.85$ MeV

(allowed)

^{80}Br + β^- $\rightarrow$ ^{80}Se

$Q = (79.918530 - 79.916522)\,u \cdot c^2 = 1.9$ MeV (allowed)

40. ^{227}Ac $\rightarrow$ ^{223}Fr + ^{4}He:

$Q = [227.027747 - 223.019731 - 4.002603]\,u \cdot c^2 = 5.0$ MeV (allowed)

^{227}Ac $\rightarrow$ ^{227}Th + β^-:

$Q = (227.027747 - 227.027699)\,u \cdot c^2 = 0.045$ MeV (allowed, barely)

^{227}Ac $\rightarrow$ ^{227}Ra + β^+:

$Q = (227.027747 - 227.029171 - 2\,(0.000549))\,u \cdot c^2 = -2.3$ MeV

(not allowed)

^{227}Ac + β^- $\rightarrow$ ^{227}Ra

$Q = (227.027747 - 227.029171)\,u \cdot c^2 = -1.3$ MeV (not allowed)

50.

$$R' = \frac{N\left(^{206}\text{Pb}\right)}{N\left(^{238}\text{U}\right)} = e^{\lambda t} - 1 = e^{\ln 2} - 1 = 1$$

where the substitution for λ occurs since the time given in the problem almost exactly matches the half-life of U-238 so that $\lambda t = \left(\ln(2)/t_{1/2}\right)t = \ln(2)$.

51. From the A values it is clear that there are $28/4 = 7$ alpha decays. Seven alpha decays reduces Z from 92 to 78, so there must be four β^- decays in order to bring Z up to 82. There are other possible combinations of beta decays (including β^+ and electron capture), but the net result must be a change of four charge units. We would have to look at a table of nuclides to determine the exact chain(s).

54. a) $t_{1/2} = 7.7 \times 10^{24}$ y $= 2.4 \times 10^{32}$ s

$$R = \lambda N = \frac{\ln 2}{t_{1/2}} \frac{N_A}{0.128 \text{ kg}} = \frac{\ln 2}{2.4 \times 10^{32} \text{ s}} \frac{6.022 \times 10^{23}}{0.128 \text{ kg}}$$

$$= 1.36 \times 10^{-8} \text{ s}^{-1} \cdot \text{kg}^{-1}$$

b)

$$\frac{10 \text{ s}^{-1}}{1.36 \times 10^{-8} \text{ s}^{-1} \cdot \text{kg}^{-1}} = 7.36 \times 10^8 \text{ kg}$$

This is not a realistic sample size.

55. Adding charge dq to a solid sphere of radius r we have an energy change

$$dE = \frac{Q \, dq}{4\pi\varepsilon_0 r}$$

where $Q = \frac{4}{3}\rho\pi r^3$ is the charge already there and ρ is the charge density. Also we know $dq = \rho dV = 4\pi\rho r^2 \, dr$.

$$dE = \frac{16\pi^2 \rho^2 r^4}{3(4\pi\varepsilon_0)} dr$$

Integrating from 0 to R we find

$$\Delta E = \int_0^R \frac{16\pi^2 \rho^2 r^4}{3(4\pi\varepsilon_0)} dr = \frac{16\pi^2 \rho^2 R^5}{15(4\pi\varepsilon_0)}$$

The charge density is $\rho = Q/V = 3Q/4\pi R^3$, so

$$\Delta E = \frac{16\pi^2 R^5}{15(4\pi\varepsilon_0)} \left(\frac{3Q}{4\pi R^3}\right)^2 = \frac{3Q^2}{5(4\pi\varepsilon_0)R} = \frac{3(Ze)^2}{5(4\pi\varepsilon_0)R}$$

60

62. **a)** Both ^{4}He and ^{16}O are "doubly magic"; that is both Z and N are magic numbers. These elements are more tightly bound than nuclei around them.

b) Because ^{208}Pb has $Z = 82$ and $N = 126$ it is doubly magic. It is particularly stable and its binding energy is high. Other nuclei around it are unstable because of the large Coulomb energy.

c) ^{40}Ca is doubly magic and is the most stable of the calcium isotopes. It is the heaviest nuclide with $Z = N$. ^{48}Ca is also stable and has $Z = 20$ and $N = 28$ which are both magic numbers. Six isotopes of calcium are stable with $Z = 20$ and various values of N.

Chapter 13

3. Probability is equal to $nt\sigma$, or

$$nt\sigma = \frac{6.022 \times 10^{23}}{238 \text{ g}} \frac{19 \text{ g}}{\text{cm}^3} (3 \text{ cm}) \left(0.6 \times 10^{-24} \text{ cm}^2\right) = 0.087$$

8. a)

$$\begin{aligned} Q &= \left[M(^{16}\text{O}) + M(^2\text{H}) - M(^4\text{He}) - M(^{14}\text{N})\right] \text{u} \cdot c^2 \\ &= 3.11 \text{ MeV (exothermic)} \end{aligned}$$

b)

$$\begin{aligned} Q &= \left[M(^{12}\text{C}) + M(^{12}\text{C}) - M(^2\text{H}) - M(^{22}\text{Na})\right] \text{u} \cdot c^2 \\ &= -7.95 \text{ MeV (endothermic)} \end{aligned}$$

c)

$$\begin{aligned} Q &= \left[M(^{23}\text{Na}) + M(^1\text{H}) - M(^{12}\text{C}) - M(^{12}\text{C})\right] \text{u} \cdot c^2 \\ &= -2.24 \text{ MeV (endothermic)} \end{aligned}$$

11. a)

$$Q = K_y + K_Y - K_x = 1.1 \text{ MeV} + 6.4 \text{ MeV} - 5.5 \text{ MeV} = 2.0 \text{ MeV}$$

b) The Q value does not change for a particular reaction.

16. a)

$$Q = \left[M(^{16}\text{O}) + M(^4\text{He}) - M(^1\text{H}) - M(^{19}\text{F})\right] \text{u} \cdot c^2 = -8.11 \text{ MeV}$$

$$K_{\text{th}} = \frac{20}{16} (8.11 \text{ MeV}) = 10.14 \text{ MeV}$$

b)

$$Q = \left[M(^{12}\text{C}) + M(^2\text{H}) - M(^3\text{He}) - M(^{11}\text{B})\right] \text{u} \cdot c^2 = -10.46 \text{ MeV}$$

$$K_{\text{th}} = \frac{14}{12} (10.46 \text{ MeV}) = 12.21 \text{ MeV}$$

18. The equation for K_{cm} is correct because we know from classical mechanics that the system is equivalent to a mass $M_x + M_X$ moving with a speed v_{cm}^2, so the center of mass kinetic energy is

$$K_{cm} = \frac{1}{2} (M_x + M_X) v_{cm}^2$$

Letting $M = M_x + M_X$ we have by conservation of momentum $Mv_{cm} = M_x v_x$, or $v_{cm} = M_x v_x / M$. Thus

$$K_{cm} = \frac{1}{2} M v_{cm}^2 = \frac{1}{2} M \frac{M_x^2}{M^2} v_x^2 = \frac{1}{2} \frac{M_x^2}{M} v_x^2$$

$$K'_{cm} = K_{lab} - K_{cm} = \frac{M_x v_x^2}{2} - \frac{1}{2} \frac{M_x^2}{M} v_x^2 = \frac{M_x v_x^2}{2} \left(1 - \frac{M_x}{M} \right)$$

$$= \frac{M_x v_x^2}{2} \left(\frac{M - M_x}{M} \right)$$

$$K'_{cm} = K_{lab} \left(\frac{M_X}{M_x + M_X} \right)$$

19.

$$K'_{cm} = \frac{M_X}{M_x + M_X} K_{lab} = \frac{14}{18} (7.7 \text{ MeV}) = 5.99 \text{ MeV}$$

$$E^* = \left[M(^{14}\text{N}) + M(^4\text{He}) - M(^{18}\text{F}) \right] u \cdot c^2 = 4.41 \text{ MeV}$$

$$E_x = E^* + K'_{cm} = 10.40 \text{ MeV}$$

22. From Chapter 12 Problem 27 we see that the mean lifetime is

$$\tau = \frac{t_{1/2}}{\ln 2} = \frac{109 \text{ ms}}{\ln 2} = 157 \text{ ms}$$

The from Equation (13.13)

$$\Gamma = \frac{\hbar}{2\tau} = \frac{6.582 \times 10^{-16} \text{ eV} \cdot \text{s}}{2 (0.157 \text{ s})} = 2.10 \times 10^{-15} \text{ eV}$$

^{17}Ne can decay by positron decay or electron capture.

27. a)

$$m = \left(4 \times 10^{-4}\right)\left(10^6 \text{ kg}\right) = 400 \text{ kg}$$

b)

$$(400 \text{ kg}) \frac{6.022 \times 10^{23}}{0.238 \text{ kg}} = 1.01 \times 10^{27} \text{ atoms}$$

c)

$$R = \left(0.69 \text{ kg}^{-1} \cdot \text{s}^{-1}\right)(400 \text{ kg}) = 276 \text{ Bq}$$

d)

$$\left(276 \text{ s}^{-1}\right) \frac{86400 \text{ s}}{d} = 2.38 \times 10^7 \text{ d}^{-1}$$

28.

$$\frac{N\left(^{235}\text{U}\right)}{N\left(^{238}\text{U}\right)} = \frac{7}{993} = \frac{N_0\left(^{235}\text{U}\right)e^{-\lambda_1 t}}{N_0\left(^{238}\text{U}\right)e^{-\lambda_2 t}}$$

where the subscripts 1 and 2 refer to the 235 and 238 isotopes, respectively.

$$\frac{N_0\left(^{235}\text{U}\right)}{N_0\left(^{238}\text{U}\right)} = \frac{7}{993} \exp\left((\lambda_1 - \lambda_2)\, t\right)$$

$$
\begin{aligned}
(\lambda_1 - \lambda_2)\, t &= \ln 2 \left(\frac{1}{t_{1/2}(235)} - \frac{1}{t_{1/2}(238)}\right) t \\
&= \ln 2 \left(\frac{1}{7.04 \times 10^8 \text{ y}} - \frac{1}{4.47 \times 10^9 \text{ y}}\right)\left(2 \times 10^9 \text{ y}\right) = 1.659
\end{aligned}
$$

Then

$$\frac{N_0\left(^{235}\text{U}\right)}{N_0\left(^{238}\text{U}\right)} = \frac{7}{993} \exp\left((\lambda_1 - \lambda_2)\, t\right) = \frac{7}{993} \exp(1.659) = 0.0370$$

which is more than five times higher than today. Natural fission reactors cannot operate because of the relatively low abundance of ^{235}U today.

31. For uranium we assume as in the previous problem that each fission produces 200 MeV of energy.

$$(1 \text{ kg}) \left(\frac{6.022 \times 10^{23} \text{ atoms}}{0.235 \text{ kg}} \right) \frac{200 \text{ MeV}}{\text{atom}} = 5.13 \times 10^{26} \text{ MeV}$$

Converting to kWh we find 2.30×10^7 kWh. On the other hand, for coal the conversion of 29000 Btu is 8.50 kWh. Therefore we see that fission produces over one million times more energy per kilogram of fuel.

32. a)

$$\frac{3}{2}kT = \frac{3}{2} \left(8.617 \times 10^{-5} \text{ eV/K}\right) (300 \text{ K}) = 3.88 \times 10^{-2} \text{ eV}$$

b)

$$\frac{3}{2}kT = \frac{3}{2} \left(8.617 \times 10^{-5} \text{ eV/K}\right) (15 \times 10^6 \text{ K}) = 1.94 \text{ keV}$$

37. a) This problem is similar to Example 13.9. First we will determine the Coulomb potential energy that must be overcome, using Equation (12.2) to determine the radius.

$$r = r_0 A^{1/3} = 1.2 \times 10^{-15} \text{ m } (12)^{1/3} = 2.7 \times 10^{-15} \text{ m}$$

Therefore the Coulomb barrier is:

$$V = \frac{q_1 q_2}{4\pi\epsilon_0 r} = \frac{\left(9 \times 10^9 \text{ N} \cdot \text{m}^2/\text{C}^2\right) (6)^2 \left(1.6 \times 10^{-19} \text{ C}\right)^2}{2.7 \times 10^{-15} \text{ m}}$$
$$= 3.1 \times 10^{-12} \text{ J}$$

We need at least this much kinetic energy to overcome the Coulomb barrier. We set this value equal to the thermal energy, namely $\frac{3}{2}kT$.

$$T = \frac{2V}{3k} = \frac{2\left(3.1 \times 10^{-12} \text{ J}\right)}{3\left(1.38 \times 10^{-23} \text{ J/K}\right)} = 1.5 \times 10^{11} \text{ K}$$

b) We use Equation (13.7) to find the Q value for the reaction:

$$Q = \left[2M(^{12}C) - M(^{24}Mg)\right] u \cdot c^2 = [2\,(12.0) - 23.985042]\, u \cdot c^2$$
$$= 13.9\,\text{MeV}$$

This includes the energy of the γ ray in the released energy.

45. a) With a half-life of 6.01 h, not much is left after one week.

b)

$$R = R_0 e^{-\lambda t} = \left(10^{11}\,\text{Bq}\right) \exp\left(-\frac{\ln 2}{6\,\text{h}}\,(216\ \text{h})\right) = 1.46\,\text{Bq}$$

c)

$$R = R_0 e^{-\lambda t} = \left(0.9 \times 10^{11}\,\text{Bq}\right) \exp\left(-\frac{\ln 2}{6\,\text{h}}\,(96\ \text{h})\right) = 1.37 \times 10^6\,\text{Bq}$$

54. a)

$$(1\ \text{kg})\,\frac{6.022 \times 10^{23}}{\text{mol}}\,\frac{1\ \text{mol}}{0.001\ \text{kg}}\,\frac{13.6\ \text{eV}}{1\ \text{atom}} = 8.19 \times 10^{27}\ \text{eV}$$

b)

$$(1\ \text{kg})\,\frac{6.022 \times 10^{23}}{\text{mol}}\,\frac{1\ \text{mol}}{0.002\ \text{kg}}\,\frac{2.22 \times 10^6\ \text{eV}}{1\ \text{atom}} = 6.68 \times 10^{32}\ \text{eV}$$

c)

$$(1\ \text{kg})\,\frac{1\ \text{u}}{1.66 \times 10^{-27}\ \text{kg}}\,\frac{931.49 \times 10^6\ \text{MeV}}{\text{u}} = 5.61 \times 10^{35}\ \text{eV}$$

55. a)

$$Q = K_{\text{out}} - K_{\text{in}} = 86.63\ \text{MeV} - 100\ \text{MeV} = -13.37\ \text{MeV}$$

This agrees very nearly with the Q computed using atomic masses (assuming all the masses are known).

$$Q = \left[M(^{18}O) + M(^{30}Si) - M(^{14}O) - M(^{34})Si\right] u \cdot c^2 = -13.27\,\text{MeV}$$

b)

$$M = -\frac{Q}{c^2} + M(^{18}O) + M(^{30}Si) - M(^{14}O) = 33.979\,\text{u}$$

which agrees very nearly with the value for the mass of ^{34}Si in Appendix 8.

59.

$$m(^{40}K) = (70\text{ kg})\,(0.0035)\,(0.00012) = 2.94 \times 10^{-5}\,\text{kg}$$

$$N = \left(2.94 \times 10^{-5}\text{ kg}\right)\frac{6.022 \times 10^{23}}{0.040\text{ kg}} = 4.43 \times 10^{20}$$

$$R = \lambda N = \frac{\ln 2}{1.28 \times 10^9\text{ y}}\frac{1\text{ y}}{3.156 \times 10^7\text{ s}}\left(4.43 \times 10^{20}\right) = 7600\,\text{Bq}$$

The beta activity is $(7600\text{ Bq})\,(0.893) = 6790\,\text{Bq}$.

Chapter 14

1. By conservation of momentum the photons must have the same energy. For each one $E = hf = mc^2$ and

$$f = \frac{mc^2}{h} = \frac{938.27 \times 10^6 \text{ eV}}{4.136 \times 10^{-15} \text{ eV} \cdot \text{s}} = 2.27 \times 10^{23} \text{ Hz}$$

2. As in the text let $mc^2 = \hbar c/R$, so

$$R = \frac{\hbar c}{mc^2} = \frac{197.3 \text{ eV} \cdot \text{nm}}{140 \times 10^6 \text{ eV}} = 1.41 \times 10^{-6} \text{ nm} = 1.41 \text{ fm}$$

4. We know that $\lambda \leq D$ and the problem says specifically to choose $\lambda = 0.1D$ with the diameter $D = 0.15$ fm. Therefore $\lambda = 0.1(1.5\,\text{fm}) = 0.15$ fm. We know from de Broglie's relationship that $p = h/\lambda$ and we can determine the kinetic energy from this momentum as follows:

$$E^2 = \left(K + mc^2\right)^2 = (pc)^2 + \left(mc^2\right)^2$$

$$K = \sqrt{(pc)^2 + (mc^2)^2} - mc^2 = \sqrt{\left(\frac{hc}{0.15 \text{ fm}}\right)^2 + (mc^2)^2} - mc^2$$

electron:

$$
\begin{aligned}
K &= \sqrt{\left(\frac{1239.8 \,\text{eV} \cdot \text{nm}}{0.15 \,\text{fm} \frac{10^{-6} \,\text{nm}}{1 \,\text{fm}}}\right)^2 + (0.511 \times 10^6 \,\text{eV})^2} - 0.511 \times 10^6 \,\text{eV} \\
&= 8.26 \,\text{GeV}
\end{aligned}
$$

proton:

$$
\begin{aligned}
K &= \sqrt{\left(\frac{1239.8 \,\text{eV} \cdot \text{nm}}{0.15 \,\text{fm} \frac{10^{-6} \,\text{nm}}{1 \,\text{fm}}}\right)^2 + (938.27 \times 10^6 \,\text{eV})^2} - 938.27 \times 10^6 \,\text{eV} \\
&= 7.38 \,\text{GeV}
\end{aligned}
$$

15. Let subscript 1 refer to the Σ, subscript 2 to the Λ, and no subscript to the photon. From conservation of momentum $| p |=| p_2 |= E/c$. From conservation of energy

$$m_1 c^2 = \sqrt{p_2^2 c^2 + (m_2 c^2)^2} + E = \sqrt{E^2 + (m_2 c^2)^2} + E$$

$$E = \frac{(m_1 c^2)^2 - (m_2 c^2)^2}{2 m_1 c^2} = \frac{(1193 \text{ MeV})^2 - (1116 \text{ MeV})^2}{2 (1193 \text{ MeV})}$$
$$= 74.5 \text{ MeV}$$

18. n (udd): $q = \frac{2e}{3} - \frac{e}{3} - \frac{e}{3} = 0$; $B = 3 \left(\frac{1}{3}\right) = 1$; $S = 3 (0) = 0$

Σ^+ (uus): $q = \frac{2e}{3} + \frac{2e}{3} - \frac{e}{3} = e$; $B = 3 \left(\frac{1}{3}\right) = 1$; $S = 0 + 0 - 1 = -1$

Λ_C^+ (udc): $q = \frac{2e}{3} - \frac{e}{3} + \frac{2e}{3} = e$; $B = 3 \left(\frac{1}{3}\right) = 1$; $S = 0 + 0 + 0 = 0$

24. We use Table 14.5 for quark properties and Table 14.4 or Table 14.6 to identify the hadrons. The spin of all quarks and antiquarks is $1/2$.

a) $c\bar{d}$; spin is 0 or 1; charge is 1; baryon number is 0; $C = 1$; $S, B, T = 0$. This is a D^+ meson.

b) uds; spin is $1/2$ or $3/2$; charge is 0; baryon number is 1; $S = -1$; $C, B, T = 0$. This could be a Σ^0 or a Λ baryon.

c) $\overline{sss}$; spin is $1/2$ or $3/2$; baryon number is -1; charge is 1; $S = 3$; $C, B, T = 0$. This is a Ω^+ baryon.

d) $\bar{c}d$; spin is 0 or 1; charge is -1; baryon number is 0; $C = -1$; $S, B, T = 0$. This is a D^- meson.

29. We begin with Equation (14.10). Since $K \gg mc^2$, this simplifies to $E_{cm} = \sqrt{2mc^2 K_{lab}}$. From the problem we know that $E_{cm} = 2K$ so $E_{cm} = 2K = \sqrt{2mc^2 K_{lab}}$. This can be rearranged to find $4K^2 = 2mc^2 K_{lab}$ or

$$K_{lab} = \frac{2K^2}{mc^2}$$

34. a)

$$p = \frac{\sqrt{E^2 - E_0^2}}{c} = \frac{\sqrt{(948.27 \text{ MeV})^2 - (938.27 \text{ MeV})^2}}{c}$$
$$= 137.35 \text{ MeV}/c$$

$$R = \frac{p}{qB} = \frac{137.35 \text{ MeV}}{(2.998 \times 10^8 \text{ m/s})(e)(1 \text{ T})} = 0.458 \text{ m}$$

b) From Equation (14.9) we have

$$f = \frac{eB}{2\pi m}\sqrt{1 - v^2/c^2} = \frac{eB}{2\pi m\gamma}$$

$$\gamma = \frac{E}{E_0} = \frac{948.27 \text{ MeV}}{938.27 \text{ MeV}} = 1.0107$$

$$f = \frac{(e)(1 \text{ T})}{2\pi (938.27 \text{ MeV})(1.0107)}(2.998 \times 10^8 \text{ m/s})^2 = 1.51 \times 10^7 \text{ Hz}$$

37. As in the previous problem

$$E_{\text{cm}} = \sqrt{(m_1 c^2 + m_2 c^2)^2 + 2Km_2 c^2} = \sqrt{(2mc^2)^2 + 2Km_2 c^2}$$

a) If $K \ll mc^2$ we neglect K, so $E_{\text{cm}} \approx 2mc^2$.

b) If $K \gg mc^2$ we neglect the first term and $E_{\text{cm}} \approx \sqrt{2Km_2 c^2} = \sqrt{2Kmc^2}$

In (a) we interpret the result to mean that at very low energies there is no extra energy available (beyond the masses of the two original particles). In (b) we see that the available center of mass energy increases only in proportion to $\sqrt{K}$, thus illustrating the great advantage of colliding beam experiments over fixed target experiments.

43. a) For a stationary target the sum of the rest energies of the products equals the total center of mass energy, so

$$E_{\text{cm}} = \sqrt{2E_0(2E_0 + K)} = (m_p + m_\Lambda + m_K)c^2$$
$$= 938 \text{ MeV} + 1116 \text{ MeV} + 494 \text{ MeV} = 2548 \text{ MeV}$$

Rearranging we have

$$\frac{E_{cm}^2}{2E_0} = 2E_0 + K$$

$$K = \frac{E_{cm}^2}{2E_0} - 2E_0 = \frac{(2548 \text{ MeV})^2}{2(938 \text{ MeV})} - 2(938 \text{ MeV}) = 1585 \text{ MeV}$$

b) In a colliding beam experiment the total momentum is zero, and we have by conservation of energy

$$2E_0 + 2K = E_0 + (m_\Lambda + m_K) c^2$$

$$K = \frac{-E_0 + (m_\Lambda + m_K) c^2}{2} = \frac{-938 \text{ MeV} + 1116 \text{ MeV} + 494 \text{ MeV}}{2}$$
$$= 336 \text{ MeV}$$

46. a) baryon number and electron lepton number not conserved

    b) not allowed - charge is not conserved

    c) allowed

    d) allowed

47. a) strangeness is not conserved

    b) charge is not conserved

    c) baryon number is not conserved

    d) strangeness is not conserved

Chapter 15

1. From Newton's second law we have for a pendulum of length L

$$F = m_G g \sin\theta = m_I a = m_I L \frac{d^2\theta}{dt^2}$$

$$\frac{d^2\theta}{dt^2} = \frac{m_G g}{m_I L} \sin\theta \approx \frac{m_G g}{m_I L}\theta$$

where we have made the small angle approximation $\sin\theta \approx \theta$. This is a simple harmonic oscillator equation with solution $\theta = \theta_0 \cos(\omega t)$ where θ_0 is the amplitude and the angular frequency is

$$\omega = \sqrt{\frac{m_G g}{m_I L}}$$

The period of oscillation is

$$T = \frac{2\pi}{\omega} = 2\pi\sqrt{\frac{m_I L}{m_G g}}$$

Therefore two masses with different ratios m_I/m_G will have different small-amplitude periods.

6. Using the mass of the neutron star and the radius given

$$\frac{\Delta f}{f} = \frac{GM}{rc^2} = \frac{\left(6.673 \times 10^{-11}\,\mathrm{m^3 \cdot kg^{-1} \cdot s^{-2}}\right)\left(5 \times 10^{30}\,\mathrm{kg}\right)}{\left(1.0 \times 10^4\,\mathrm{m}\right)\left(2.998 \times 10^8\,\mathrm{m/s}\right)^2}$$

$$= 3.712 \times 10^{-1}$$

The wavelength is affected by the same factor, so the redshift at the given wavelength is

$$\Delta\lambda = \left(3.712 \times 10^{-1}\right)(550\,\mathrm{nm}) = 204\,\mathrm{nm}$$

11.

$$r_s = \frac{2GM}{c^2} = \frac{2\left(6.673 \times 10^{-11}\,\mathrm{m^3 \cdot kg^{-1} \cdot s^{-2}}\right)\left(1.90 \times 10^{27}\,\mathrm{kg}\right)}{\left(2.998 \times 10^8\,\mathrm{m/s}\right)^2}$$

$$= 2.82\,\mathrm{m}$$

12. From Equation (15.7) we have

$$T = \frac{\hbar c^3}{8\pi k G M}$$

The right-hand fraction equals

$$\frac{(1.0546 \times 10^{-34} \text{ J} \cdot \text{s})(2.998 \times 10^8 \text{ m/s})^3}{8\pi (1.381 \times 10^{-23} \text{ J/K})(6.673 \times 10^{-11} \text{ m}^3 \cdot \text{kg}^{-1} \cdot \text{s}^{-2})(1.99 \times 10^{30} \text{ kg})}$$

and so

$$T = 6.17 \times 10^{-8} \text{ K}$$

15. a) We use the results of Example 15.3 that give the time in terms of the mass. Rearranging the equation, using the age of the universe as 13.7 billion years, and using the value of α from problem 16, we have

$$
\begin{aligned}
M_0 &= (3\alpha t)^{1/3} \\
&= \left[3 \left(3.965 \times 10^{15} \text{ kg}^3/\text{s} \right) \left(13.7 \times 10^9 \text{ y} \frac{3.156 \times 10^7 \text{ s}}{1 \text{ y}} \right) \right]^{1/3} \\
&= 1.73 \times 10^{11} \text{ kg}
\end{aligned}
$$

b) Current evidence for the smallest black holes require a mass of about 5 to 20 times the mass of the Sun. The mass from part a) is only $M_0 \approx 10^{-19} M_{\text{Sun}}$ which is too small to form a black hole.

18. Set the change in the photon's energy equal to the change in gravitational potential energy:

$$\Delta E = h\,\Delta f = -\frac{GMm}{r_1} - \left(-\frac{GMm}{r_2} \right) = -GMm\left(\frac{1}{r_1} - \frac{1}{r_2} \right)$$

where M is the mass of the earth and m is the equivalent mass of the photon. Now $m = E/c^2 = hf/c^2$, so

$$h\,\Delta f = -\frac{GMhf}{c^2}\left(\frac{1}{r_1} - \frac{1}{r_2} \right)$$

$$\frac{\Delta f}{f} = -\frac{GM}{c^2}\left(\frac{1}{r_1} - \frac{1}{r_2} \right)$$

Chapter 16

1.

$$1 \text{ pc} = \frac{1 \text{ au}}{\tan 1''} \left(1.496 \times 10^{11} \text{ m/au}\right) = 3.086 \times 10^{16} \text{ m}$$

$$1 \text{ ly} = \left(2.9979 \times 10^8 \text{ m/s}\right) \left(365.25 \text{ d/y}\right) \left(86400 \text{ s/d}\right)$$
$$= 9.461 \times 10^{15} \text{ m}$$

$$1 \text{ pc} = \frac{3.086 \times 10^{16} \text{ m}}{9.461 \times 10^{15} \text{ m/ly}} = 3.26 \text{ ly}$$

7. The π^+ ($E_0 = 140$ MeV) is more massive than the π^0 ($E_0 = 135$ MeV), so the π^+ was formed first. With $\Delta mc^2 = k\Delta T$ we have

$$\Delta T = \frac{\Delta mc^2}{k} = \frac{5 \text{ MeV}}{8.617 \times 10^{-11} \text{ MeV/K}} = 5.80 \times 10^{10} \text{ K}$$

8. Set the deuteron binding energy 2.22 MeV equal to kT:

$$T = \frac{2.22 \text{ MeV}}{k} = \frac{2.22 \text{ MeV}}{8.617 \times 10^{-11} \text{ MeV/K}} = 2.58 \times 10^{10} \text{ K}$$

14. We begin with Equation (16.1).

$$v = HR = \frac{71 \text{ km/s}}{\text{Mpc}} \left(4 \times 10^9 \text{ ly}\right) \left(\frac{1 \text{ Mpc}}{3.26 \times 10^6 \text{ ly}}\right)$$
$$= 8.71 \times 10^4 \text{ km/s} = 0.29 \, c$$

15.

$$R = \frac{v}{H} = \frac{15000 \text{ km/s}}{71 \text{ km/s/Mpc}} = 211 \text{ Mpc or about } 689 \text{ Mly}$$

17. a) From Equation (16.18) we see that for a redshift of 10 we have $\Delta\lambda/\lambda_0 = 10$, so

$$10 = \sqrt{\frac{1+\beta}{1-\beta}} - 1$$

74

which can be solved to find $\beta = 0.984$ or $v = 0.984\,c$.

b)

$$R = \frac{v}{H} = \frac{0.984\,(299790\,\text{km/s})}{71\,\text{km/s/Mpc}} = 4155\,\text{Mpc}$$

$$4155\,\text{Mpc}\left(\frac{3.26\,\text{Mly}}{1\,\text{Mpc}}\right) = 13.5\,\text{Gly}$$

23. For $H = 55$ km/s/Mpc we have

$$H = (55000\ \text{m/s/Mpc})\left(\frac{1\,\text{Mpc}}{3.086 \times 10^{22}\,\text{m}}\right) = 1.78 \times 10^{-18}\ \text{s}^{-1}$$

$$\rho_c = \frac{3H^2}{8\pi G} = \frac{3\left(1.78 \times 10^{-18}\ \text{s}^{-1}\right)^2}{8\pi\left(6.67 \times 10^{-11}\ \text{m}^3 \cdot \text{kg}^{-1} \cdot \text{s}^{-2}\right)} = 5.67 \times 10^{-27}\ \text{kg/m}^3$$

For $H = 85$ km/s/Mpc we have

$$H = (85000\ \text{m/s/Mpc})\left(\frac{1\,\text{Mpc}}{3.086 \times 10^{22}\,\text{m}}\right) = 2.75 \times 10^{-18}\ \text{s}^{-1}$$

$$\rho_c = \frac{3H^2}{8\pi G} = \frac{3\left(2.75 \times 10^{-18}\ \text{s}^{-1}\right)^2}{8\pi\left(6.67 \times 10^{-11}\ \text{m}^3 \cdot \text{kg}^{-1} \cdot \text{s}^{-2}\right)} = 1.35 \times 10^{-26}\ \text{kg/m}^3$$

33.

$$\text{Redshift} = \frac{\Delta\lambda}{\lambda_0} = \frac{582.5\ \text{nm} - 121.6\ \text{nm}}{121.6\ \text{nm}} = 3.79$$

$$3.79 = \sqrt{\frac{1+\beta}{1-\beta}} - 1$$

so $\beta = 0.92$ and $v = 0.92c$.

34. From the Doppler effect we know

$$\frac{\lambda}{\lambda_0} = \sqrt{\frac{1+\beta}{1-\beta}} = \frac{\lambda_0 + \Delta\lambda}{\lambda_0} = 1 + \frac{\Delta\lambda}{\lambda_0} = 1 + z$$

38. We begin with Equation (16.21):

$$\rho_c = \frac{3H^2}{8\pi G}$$

$$
\begin{aligned}
\rho_c &= \left(\frac{H \text{ in } (\text{km/s})/\text{Mpc}}{100}\right)^2 \frac{3 \times 10^4}{8\pi \left(6.6726 \times 10^{-11} \text{ m}^3 \cdot \text{kg}^{-1} \cdot \text{s}^{-2}\right)} \\
&= \left(\frac{H}{100}\right)^2 \left(1.789 \times 10^{13} \text{ km}^2 \cdot \text{s}^{-2} \cdot \text{Mpc}^{-2} \cdot \text{m}^{-3} \cdot \text{kg}^{-1} \cdot \text{s}^2\right)
\end{aligned}
$$

We multiply this expression by the following conversion factors:

$$\left(\frac{\text{Mpc}}{3.086 \times 10^{22} \text{ m}}\right)^2 \left(10^3 \frac{\text{m}}{\text{km}}\right)^2$$

So

$$\rho_c = \left(\frac{H}{100}\right)^2 \left(1.88 \times 10^{-26} \text{ kg/m}^3\right) = \left(\frac{H}{100}\right)^2 \left(1.88 \times 10^{-29} \text{ g/m}^3\right)$$